THE ALIEN BIKER CHICK

ISBN 978-1-946886-23-1

This book is a work of fiction. References to real people, events, establishments, organizations, or locales are intended only to provide a sense of authenticity, and are used fictitously. All other characters, and all incidents and dialogue, are drawn from the author's imagination and are not to be construed as real.

Published by
Middle River Press
middleriverpress.com

First Printing

Printed in the USA

THE ALIEN BIKER CHICK

KEN KAYE

Middle River Press

1

For most folks in the horsey town of Davie, Florida, it was a spectacular mid-February day, graced with a mild breeze, mellow sun and a dazzling blue sky.

For George Leon, it may as well have been raining frogs. Sitting alone at a picnic table in Tree Tops Park, he was gorging on a meatball sub, pissed off at the world.

His editor had given him another stupid assignment, a gooey feature about a new petting zoo at the rodeo grounds. He would have much preferred to cover a triple homicide.

His no-good brother, Mickey, was begging for bail money again after another barroom brawl. His neighbor's arrogant French Poodle had relieved itself in his backyard again.

And, worst of all, after six good months together, his girlfriend had dumped him by way of a text just that morning.

We're thru. It's time for me to find a real soulmate.

That really frosted his ass. The woman should have had the *guts* to tell him in person why she wanted to break up. Hurt and humiliated, George – Geo to his friends and coworkers – felt like the planets and stars were lined up against him.

Out of frustration, he gnawed off a huge chunk of the sub, stuffing so much meat, cheese and bread into his maw that his cheeks swelled. Tomato sauce oozed out of the corners of his mouth.

He closed his eyes, trying to find some semblance of peace, even as he ate. At least he had come to the right place. Tree Tops Park was full of leafy trees, serene lakes and green fields. He often came here during his lunch hour to seek sanctuary from the chaos of South Florida and the bullshit at his office.

All of a sudden, he heard a motorcycle racing on a nearby roadway.

From the crescendo of its roar, he could tell it was approaching the park entrance not far from his picnic table, in a grassy field. Eyes popping open, he stopped chewing and remained still, sensing danger.

Seconds later a mean-looking bike charged up to him and slid to a halt, its tires leaving divots in the grass, gray smoke wafting from its cylinders. Placing feet on the ground, the rider shut down the engine, which continued ticking in the cool air.

Geo didn't know much about motorcycles. But he could tell this thing wasn't some motor scooter intended for Sunday outings. It was big and nasty, something out of a Hells Angels movie.

Dismounting, the rider pulled off a black helmet, revealing a woman he had never seen before. Appearing to be Hispanic, she shook her head, allowing long, brown hair to cascade over her shoulders.

Her well-worn jeans were tight; her knee-high, black boots were equipped with thick, three-inch heels. Her leather jacket creaked as she placed the helmet on the back of the bike.

She looked around as though checking if someone had followed her and sat down across from him at the picnic table. She stuck out a hand, her fingers long and slender, her nails short and dirty.

He thought she wanted to shake hands. Instead, she demanded, "Give me half your sandwich."

Mouth still full, Geo slid the other half of the meatball sub in its wrapper papers over to her. She took a hefty bite, her eyes expressionless like a lioness gnawing on raw meat. He resumed chewing until she caught him staring.

"The hell you looking at?" A slight Spanish accent all but confirmed she was a Latina.

Geo wiped his mouth with a napkin. Always the professional, he was wearing a white dress shirt, a light-blue tie, beige slacks and brown leather shoes. The matching suit jacket was hanging in the Jeep.

"Sorry. I didn't mean to offend you."

Why are you here? You want money? Or do you randomly steal people's lunch?

Rather than seek answers, he decided to let this play out.

She continued working on the sub, again glancing around. Surveying the area, he saw no one in the vicinity, nearby picnic tables empty. He picked up his large Diet Coke, sipped through a straw and subtly studied her. As a newspaper reporter he was trained to be observant.

Based on her defiant demeanor, he guessed she was a biker chick who didn't take shit from anyone. He further speculated that she was from someplace like Puerto Rico or Cuba.

In any case, this much was clear: she was hot. Her mouth was wide, her lips full. She possessed olive skin, an hour-glass figure and a beautiful jawline. Her eyes were so dark brown they were almost black.

A good-looking man himself, on the tall side with sandy-brown, neatly trimmed hair, Geo had been with several gorgeous women, including a supermodel, an executive secretary and a couple of television personalities.

But he had never connected with a biker chick, one of his secret fantasies. Was this his chance? Maybe. Then again, after the girlfriend had stomped on his heart, he wasn't feeling very sociable.

Without asking, the biker chick grabbed his Diet Coke and took a long pull through his straw. She eyed what was left of his sub, which he had placed on his napkin. "You going to eat that?"

He gave it to her. Three bites later she was done. She yet again looked around to make sure the coast was clear. He wanted to ask, *Who's chasing you? The cops? You rob a bank?* He again remained quiet.

"You horny?" she finally asked. "You want to take me right here on this table?"

He stared at her for a moment and grinned. He guessed this rough-and-tumble babe was trying to maneuver him into a vulnerable position, like with his pants around his ankles, so that she could steal his wallet and watch. Maybe she had an accomplice-boyfriend hiding in the bushes, ready jump in on the ambush.

"The fuck you smiling at? I asked you a question, pretty boy."

"Actually, you asked two questions. But I'm not buying whatever you're selling, ma'am."

"I look like a *ma'am* to you, pretty boy? I ain't selling nothing. You want to fuck me or not?"

He considered another possibility: she was mentally unstable. "Who you running from?"

"What makes you think I'm running?"

"You keep looking over your shoulder. You do something illegal? Break the law?"

"None of your fuckin' business, pretty boy."

Although she had a command of the English language, he guessed her Spanish was much better. He admired anyone who was bilingual considering he had tried hard to learn Spanish but couldn't pick it up.

Suddenly the deep roar of distant motorcycles filled the air. The volume of the powerful engines increased rapidly, indicating a large pack was closing in fast.

She jumped up from the picnic table. "Sorry, Mr. Clean. Gotta go. Thanks for the sub."

Geo furrowed his brow. "You're running from bikers?"

Much to his surprise, she came around to his side of the table, lightly cupped a hand over his crotch and gently kissed him. Her lips tasted at the same time sweet and salty, immediately arousing him.

"Meet me for dinner tonight, pretty boy. I'll tell you all about it. Shorty's, eight o'clock."

She gave his crotch a light squeeze before scrambling to her bike, pulling on the helmet and hopping on. She kick-started the engine, gave him a mock military salute and charged toward the park entrance. He saw her make a right turn on Nob Hill Road.

A minute later more than a dozen bikers streamed onto the park's internal road. To a man, they wore Nazi stormtrooper helmets, dark sunglasses, black leathers and chains. Two of them cut away from the pack and rolled up to his table.

"You see a Mexican chick on a hog come through here?" one yelled at Geo over the din of his bike's high-idling engine.

"Sure haven't."

"Fuck," the biker said and turned to his buddy. "She must have a hightailed it out of here. Let's go."

The two bikers revved their engines menacingly, the volume as loud as a jet engine, and peeled away to join the rest of the gang. Geo noted the name emblazoned on the backs of their jackets: The Vicious Varmints, a notoriously evil bike club. The bikers exited the park en masse and turned north on Nob Hill.

Geo threw his lunch papers into a nearby trash bin, walked to the other side of the picnic table and saw the tire tracks the biker chick had left in the grass. He was glad the two goons weren't very observant; otherwise they would have realized she had been right there.

Clearly the woman had done something to rile the biker gang. Geo thought about trying to follow her in case she needed to be rescued. But no. She and her pursuers were speeding away far too fast. All he could do was hope she could escape from them.

He climbed into his cream-colored Jeep Wrangler Sahara and headed back to work.

2

As he drove east on Griffin Road toward his office, his mind swirled around that crazy encounter. A hot *chica* drops out of nowhere, snares his lunch and asks if he wants to get it on right there on the picnic table. Then she cups his crotch and asks to meet for dinner.

What the hell?

He again guessed she was up to no good. She had probably spotted a convenient mark and planned to rob him. Then the bikers showed up and spoiled her fun. She had likely pilfered something from them, prompting the chase.

In any case, the biker chick might have been outright mouthwatering and represented one of his most lustful fantasies. But she was obviously trouble. He decided it would be wise to pass on her dinner invitation.

Dark aviator sunglasses snug against his face, he saw two horseback riders crossing the road a quarter of a mile ahead, a common sight in Davie. Though a suburb of metropolitan Fort Lauderdale, the town had a country feel with cattle pastures, grassy fields and nurseries.

Horses, however, were its main claim to fame. It was home to several stables and a network of riding trails. Hitching posts, allowing riders to tether their steeds, could be found in front of restaurants, bars and other businesses. The rodeo grounds, near Davie Town Hall, were a popular attraction.

Geo stopped to let the riders pass in front of his vehicle. They appeared to be a mother and teen-age daughter, both wearing cowboy hats. He waved at them as their horses clopped to the other side of the roadway. They waved back with grateful smiles.

At Davie Road, he turned into the downtown area and a half a mile later parked in front of the *West Broward Banner*. The small newspaper published twice a week, on Tuesdays and Saturdays, and circulated throughout western Broward County. Its modest stand-alone brick building, at one time a bank branch, was wedged between a McDonald's and the Beer Barn, a local watering hole.

He grabbed his suit jacket, hanging behind his driver's seat, and stepped inside. He nodded at the pretty young receptionist, who was stationed near the front door, and went to his desk at the back of the newsroom. Along the way, he passed two women reporters working at their computers.

"Good afternoon, ladies."

Both smirked. Geo knew the other reporters made fun of him because he always dressed in jackets and ties, while they usually dressed casually. He didn't care. From years of experience, including a long stint at a big-city newspaper, he knew appearances were important.

Tossing his keys and sunglasses onto a gray-metal desk, he hung his jacket on the back of his chair and flopped down. He gazed out the picture window next to his desk. It provided a beautiful view of the McDonald's drive-thru, Moe's Garage across the street and traffic on Davie Road.

It was going to be a long afternoon. The last thing he wanted to do was write a silly feature about the petting zoo, particularly after receiving that text from his now *ex*-girlfriend.

It's time for me to find a real soulmate.

How she could be so heartless? Whatever the reason, he knew he had to put her out of mind and focus on work, the professional thing to do. He prided himself on being professional.

Swiveling to his computer, he prepared to start writing about how little kids enjoyed playing with baby goats and ponies. Then Bud Kowalski called on his cellphone.

"You ain't going to believe this. A space shuttle landed at a remote airport in the Everglades last night. Most amazing thing I've ever seen. We don't know where it came from or who it belongs to."

Geo sat up straight, sensing a big story. Kowalski was a field agent for the Federal Aviation Administration, specializing in security issues. He was responsible for investigating everything from drug smuggling operations to airliner crashes.

He also was a trusted source, someone who really knew his stuff. The two had worked closely on several aviation stories over the years and developed a mutual respect.

"A space shuttle? With stubby wings and huge engines?"

"Yep, a fricking space shuttle," Kowalski confirmed. "Only this thing isn't American. It looks suspiciously like next-generation Russian."

"I didn't know the Russians developed a space shuttle."

"We didn't either. We thought their shuttle program failed. But the Russians are always doing shit behind our backs."

Geo grabbed a notepad and pen. "You say this shuttle is at an airport in the Everglades?"

"Dade-Collier Training and Transition Airport. It's only got one runway. But it's thirteen-thousand feet long, long enough to handle a shuttle."

"And you have no idea how it got there?

"It's a major-league mystery. It left no radar signature. So, unless someone towed the damn thing over there on a truck, it must have flown at treetop level for miles below radar. But we haven't received any reports about low-flying aircraft."

"Could it have been towed?"

"Not likely. It's too big to get under the bridges and past all the powerline poles along the roads leading to that airport. On top of that, its engines were searing hot when we inspected it this morning. That indicates it landed shortly before we found it."

Geo jotted that down. "What about the airport employees? Didn't they see it fly in?"

"The airport is so slow that only one person works there overnight, and that person was sound asleep in a double-wide trailer. We talked to the worker. He didn't hear or see a thing."

"Jesus."

"What's even more weird, the spacecraft is locked tight and we can't find the pilots, or cosmonauts in the event it's Russian. We don't think they're still inside. We looked through front windows and the spacecraft appeared to be empty."

"No pilots? Then what happened to them?"

"Either they fled shortly after landing or the thing is a drone without pilots. One way or the other, we got the cops looking for any crew members."

Geo took down more notes. "What's your theory, Bud?"

"We figure it suffered some sort of malfunction while in space and landed here either by mistake or out of emergency. We think it approached from the Atlantic, keeping its radar profile to a minimum. And, knowing the Russians, the Kremlin probably ordered the pilots, if any were on board, to disappear to avoid being questioned."

"Wow."

"The FAA and NSA are holding a joint press conference at the airport at three o'clock this afternoon. All the major networks will be here. This is a pretty big deal, Geo, so I thought you might like to attend. This is your kind of story."

It sure the hell is, Geo thought. It was a story that held potential to jumpstart his career, propel him back to a real newspaper again or maybe even a television station.

"Thanks for the heads-up, Bud. I'll be there if I can."

3

Geo jumped up, eager to tell his editor, Samantha Pauline Vandermark, "Sam" to her staff, about this blockbuster story. Her office was in the back of the newsroom, allowing her to monitor her employees through a large window.

On this Monday afternoon, he found her leafing through *Vogue.* As usual, she was primped to the max, blue eyeshade and shimmering red lipstick. Her shoulder-length, blond hair was perfectly coiffed. Her red pantsuit was Anne Taylor, embellished with oversized gold-loop earrings.

Geo figured the runway model look was an attempt to portray herself as a skilled supervisor, though he considered her anything but.

"Got one hell of a story, Sam," he said, sitting in a cushioned chair in front of her desk. While her office was furnished with modern furniture and plush blue carpeting, the rest of the newsroom resembled a Salvation Army outlet store, cheap desks and a worn-tile floor.

She glanced at him over her designer reading glasses, which Geo guessed were for show, lacking corrective lenses. "What's the status of the petting zoo?"

Screw the fricking petting zoo.

"It's coming along. Sam, the FAA tells me a space shuttle, possibly Russian in origin, landed at this remote airport in the Everglades. They have no idea how it got there and they can't find any pilots. It's a big deal, right in our backyard."

Sam put down her magazine. Her expensive, French perfume wafted in his direction. "That's nice. But I need the petting zoo story this afternoon. It's going on the cover of the next edition. Hope you got some cute photos of kids playing with animals."

He wanted to lunge over the desk and throttle her. Her complete lack of news judgement was infuriating.

"Sam, you'll have the petting zoo story. The feds are holding a presser in a couple of hours. I think we should be there. This could be the biggest story this little paper will ever publish. The *Banner* could sure use some hard news."

She shook her head, the gold-loop earrings dangling. "We don't do hard news, and the Everglades are hardly in our backyard. I need you to finish the petting zoo story pronto because your next assignment is about a local woman who won a chocolate-chip-cookies baking contest."

His jaw dropped. She deemed a story about chocolate chip cookies to be more important than a rogue space shuttle under investigation by the feds?

Are you fricking kidding me?

On the other hand, he wasn't that surprised. Sam wouldn't know a big news story if it walked up and bit her on the ass.

"Sam, I'll do whatever you need. But I'm begging you. Let me attend the presser. Let's get on record with this shuttle thing. Otherwise, we'll look stupid if we ignore it."

The moment he said it, he regretted it. It was tantamount to calling Sam stupid. Her eyes narrowed.

"I know you used to be a bigtime reporter for a big-city newspaper, Geo. But you're here now, and this is not the *New York Times.*"

"I worked for the *Boston Globe*, Sam."

"Whatever. Our readers don't want stories about murder or mysterious space shuttles. They can see that stuff on TV. Now, you knew the deal when I hired you."

"I did, but –"

"But nothing," she said, getting up and grabbing her Louis Vuitton purse. "I'm going to the nail salon. Make sure the petting zoo story is done before I get back."

He knew the deal. The *West Broward Banner* only published happy, intensely local news. It would blow up an elementary school softball game into a front-page story with a bold headline and colorful photos.

It did the same with garage sales and new store openings. When the local Burger King started offering vegetarian Whoppers, it gave the story a huge headline more appropriate for a presidential assassination.

Yet the small newspaper raked in a ton of money. It was chock-full of ads from businesses throughout South Florida thanks to its local flavor. Meanwhile, newspapers across the nation were dying, victims of the digital age.

Contributing to its success, the *Banner* profited from a low overhead. It had only four reporters, three of whom were fresh out of college. At thirty-three, Geo was the "old man" of the staff, the only one with any real experience.

Otherwise, the editorial department included the receptionist, a copy editor who doubled as the online producer, and Sam, who, in her mid-twenties, was a recent college graduate herself.

While the other reporters had no qualms about Sam, Geo knew she was incompetent. Her editing skills were weak and she consistently made bad management decisions.

He also knew how she landed the job. She was an attractive woman who was sleeping with the publisher, Joe Cronk, a retired corporate executive, who happened to be married.

Late one Sunday night, Geo had gone into the newsroom to do some work and spotted Sam and Cronk going at it on her office couch. Unsure if he had been spotted, Geo quietly left, never making mention to Sam or anyone else what he witnessed.

Though he held her skills and morals in low regard, Geo was indebted to Sam; she had offered him a job when no other newspaper would even grant him an interview.

He had been a top investigative reporter at the *Boston Globe* for seven years, covering terrorist bombings, banking scandals and airline disasters. Along the way, he had won several journalistic awards and built a sterling reputation.

Then, for survival reasons, *Globe* began cutting staff, notably its highest paid reporters, and Geo was one of them. Rather than risk being laid off, he had accepted a job at the *Universal Planet*, a trashy supermarket tabloid based in South Florida.

The *Planet* paid him handsomely and allowed him to escape harsh Boston winters in favor of balmy warmth. Initially, it was fun to write fictious stories, such as one claiming that Donald Trump and Hillary Clinton were secret lovers.

But Geo soon grew tired of producing nonsense. At heart, he was a hardcore reporter who wanted to make a difference by identifying bad guys, helping vulnerable people and righting the world's wrongs.

After applying to several daily newspapers and attempting to break back into mainstream journalism, he found he had lost credibility by working for a tabloid.

"I'd rather hire a janitor than someone who's worked for one of those rags." one editor told him.

Fearing his reputation had been permanently tainted, he resigned from the *Universal Planet* and almost gave up on journalism altogether. Then he saw the *West Broward Banner* was looking for a general assignment reporter. Though basically a shopper, it still offered the opportunity to redeem his integrity.

During his interview, Sam had suggested that he was too experienced for the position. But he had insisted that he would have no problem working for a small paper and writing soft features. "Journalism is journalism," he had said.

Sam hired him on the spot. The pay wasn't great, but it was enough to make payments on his Jeep and townhouse.

Now six months into the job, Geo resolved to stay put in hopes of rebuilding his reputation and eventually returning to a metropolitan newspaper. If that meant putting up with Sam and her dumb assignments, so be it.

Yet he prayed that a major story would fall into his lap, and now one just did, this strange space shuttle. One way or the other, he was determined to write about it whether Sam liked it or not.

4

He was halfway through writing the petting zoo story when Katrina Tucker called on his desk phone. He pulled on his phone headset and reluctantly answered.

"Hey, Geo. What's up?"

What's up? You dumped my ass four hours ago via text, about the coldest way to say goodbye. That's what's up, you lousy bitch.

"Hi Katrina. Not much."

Why was she calling? To confirm he got the text? To make sure he knew they were over?

"You having a good day?" she asked as though nothing had changed.

Yeah. Just like JFK had a good day at Dealey Plaza.

If they were still together, chatting confidentially, he might believe she was genuinely interested in how he was doing. But not now.

"Look, Katrina. I got the text. Don't worry. I'm not going to stalk you or beg you to take me back. I'll move along."

"I'm calling because I want you to understand why I sent the text."

"Doesn't matter why. I get it. You want me to get lost. I respect that. Okay? That make you feel better? Have a good life, Katrina."

"I still care about you, Geo," she said before he could hang up. "You need to know that."

"That's nice, Katrina. But sometimes things just aren't meant to be. I gotta go. See you – "

"Want to get together this afternoon? I'd like to talk about all this face to face."

He drummed his fingers on his desk. What was there to talk about? Her text had bluntly declared, *We're thru.* Did she want to break up or not?

"Just say what you have to say, Katrina."

"I want to make sure you don't hate me."

Hate her? He was madly in love with her. At six-foot-three, she had been an All-American basketball player at the University of Florida. Now a broadcaster for a sports television network, she was a sinewy, statuesque woman with glorious brown eyes and a ravishing face.

Beyond her beauty and athleticism, he loved her smile and zest for fun. In their six months together, they had dined and danced at nightclubs, splashed in the ocean and huffed and puffed through salty afternoons.

Until she sent that text, he was contemplating marriage. Then she goes and fractures their special bond. It hurt just to hear her voice.

"You said it's time you look for a real soulmate in that text, Katrina. I thought I was your soulmate."

"I need to explain exactly what I meant by that."

"Please do."

"Geo, you're a beautiful man. But you're a white man, and I'm a black woman. In case you haven't noticed, a very black woman. I'm proud of that. I'm proud of my Nigerian heritage. I believe that I should honor that heritage by marrying someone who can give me a black baby. That's not you."

He was speechless, slashed even deeper. Yet he had no choice but to respect her conviction.

"I understand, Katrina. You have every right to feel that way."

"Thank you, Geo. But that doesn't mean I don't have feelings for you, because I do. I still want you. That's why I'm asking. Want to get together this afternoon? One last time?"

After breaking up with him that morning, she wanted to screw in the afternoon? Though he still craved her, he knew it would sting badly to part ways with her afterward.

"I don't think so, Katrina. It's time for both of us to move along. Goodbye."

He hung up, took off his phone headset and pinched his eyes shut.

Not far away, the two women reporters were clacking on their keyboards. The online producer was eating a late lunch at her desk. He was in hell.

The fact that Katrina was leaving him because of his skin color seemed so unfair that his gut ached. To him, people were people, no matter their race, religion or origin. Yet Katrina couldn't accept an interracial union, apparently feeling it would be unnatural.

Why did she connect with me in the first place?

All he could do now was try to forget the previous six months had ever happened. That would be difficult, if not impossible. But maybe in time he would find someone new, someone who would accept him for—

His eyes popped open. He thought about the biker chick. She had invited him to meet her for dinner that night. Although he had been inclined to steer clear, he decided to take her up on her offer, see if he could get a better read on her. Though she seemed a dangerous proposition, she was also a stone-cold fox.

He turned in the petting zoo story at two-thirty. That didn't leave much time to attend the press conference at the Everglades airport but he still wanted to try.

As Sam was still at the nail salon, he sent her an email saying he was leaving the office early to cover the presser and would make up the time later. That, no doubt, would piss her off.

But he didn't care. His journalistic juices were flowing and he wasn't about to let an incompetent editor, who placed no importance on a potentially huge, international story, stop him.

Jumping in his Jeep, he charged toward southwest Miami-Dade County, winding through scrublands and open grasslands. Along the way, he listened to the sound of the big tires churning over the roadways rather than the radio.

Geo loved driving the Jeep Sahara, a big SUV that possessed many of the same pragmatic characteristics of a military Humvee. His previous vehicle had been a Porsche 911, which cost more than a hundred grand.

That was when he was making gobs of money at the *Universal Planet* and could afford to be vain. Though he loved the Porsche too, he gladly traded it in on the Jeep after starting at the *Banner*, preferring to be practical rather than macho.

When he arrived at Dade-Collier Training and Transition Airport, a parade of news vans was departing. The press conference was already over. Disappointed, Geo parked on grass near a large tent, which served as a command post.

The shuttle sat on a concrete ramp about fifty yards away, numerous officials and military personnel milling around it. Hopping out of his Jeep, Geo walked closer to the spacecraft and studied it.

Though smaller than an American shuttle, it was still sizeable. Each of its stubby wings had oversized, turned-up winglets. In addition to the front windshield, it had one window on each side behind the cockpit and was painted dark green. It otherwise had no markings whatsoever.

"Strange looking thing, isn't it?"

Geo turned to see Bud Kowalski, the FAA field agent, standing next to him.

"Hey, Bud. Came as soon as I could. But I guess the presser is over."

"It didn't last long."

"Can you fill me in?"

"No prob. Come on, let's take closer look."

They walked over to the spacecraft. Kowalski showed his FAA badge to a military guard posted near its nose.

"Please don't touch the ship," the guard said.

"Don't worry. We won't," Kowalski assured.

Geo examined the shuttle's exterior surfaces. "Looks a bit dirty. But there's not a dent or a scratch, no evidence that it's been flying through space."

Kowalski nodded. "It's probably got a hardened, weather-resistant, asteroid-resistant surface."

"You said this thing is Russian?"

Kowalski fished a cigarette out of a pack of Marlboros and lit up with a cheap lighter. His hair was thick and gray, his face tan and weather-worn from too many hours in the sun. The back of his blue jumpsuit was emblazoned with the yellow letters, "FAA."

He blew a plume of smoke upwards. "That's our thinking at the moment. But that's not confirmed. As we told the media at the press conference, the Russians refuse to comment, and our satellites never picked up the heat signature of a launch from Russia."

"Then why do you think it's Russian?"

"Process of elimination. Other than the United States, Russia is the only other nation which has built shuttles. The Chinese claim to be developing a reusable space shuttle, but that's probably bullshit. For now, they're more interested in putting people on the moon."

"But you said there's no evidence this thing blasted off from Russia."

"Correct. But it might have been launched from one of Russia's territories, and there are lots of those."

"Ah."

"It's possible that a private concern like SpaceX built it. But if that was the case, they surely would have let the US government know considering they work closely with NASA and need government approvals."

Geo nodded. "Are you still looking for the crew?"

"We're becoming more convinced no one was on board. Though we saw something that looked like a bed inside the craft, we're not sure what it is. Otherwise, we didn't see evidence of crew activity."

"Did any surveillance cameras around the airport spot it landing?"

"Nope. Most airports, this one included, don't invest in surveillance cameras for runway and ramp areas. They spend a lot of money on other security measures."

"Security cameras sure would have helped in this case."

"No argument here. At this point, the best guess is that this thing is a big sophisticated drone. Something went wrong while it was in space and its controllers, probably on the ground somewhere in Russia, were forced to make an emergency landing here."

"What's the next step?"

"We're going to try to get inside the thing. But like I told you, it's locked super tight. We have to figure out a way to gain entry."

"Was all of this covered at the presser?"

Kowalski dropped his cigarette butt to the concrete and ground it out with a black work shoe. "Yep. You're pretty much up to speed."

"Okay. Thanks, Bud. Much appreciate your –"

"Here's what we didn't tell the other media, and it's yours exclusively, if you're interested."

Geo pulled out his notebook and pen. "Of course, I'm interested."

"We think this spacecraft might be intended for military purposes, most likely to spy on us. In a worst-case scenario it might have attack capability, using missiles, lasers or some other advanced weaponry, from outer space."

"Good Christ. The Russians would attack us from space?"

"That, unfortunately, is a distinct possibility. They might have been working on a secret shuttle program explicitly for military purposes. Their previous shuttle program, for space exploration, was scuttled back in the eighties."

"Why do you think it's for military use?"

"Mainly because no one detected it arriving on radar or by satellite. So, it obviously has great stealth characteristics. It's also likely a drone, and it's much safer to attack a target with a drone than a manned craft."

"Jesus. But outer space as a battleground? Isn't that kind of farfetched?

"Hardly. In 1985, during the Ronald Reagan era, the Department of Defense created the US Space Command to give our armed forces a military advantage from space. It was disbanded in 2002 after the Nine-Eleven terrorist attacks, but the Trump Administration reactivated it in 2018. Trump went a step farther in 2019, converting the Space Command into the US Space Force, today the sixth branch of the US military."

"Didn't know that," Geo confessed.

"The Russians probably saw the Space Force as a threat. Maybe they developed this shuttle in response. The Russkies don't stand around holding their dicks when it comes to an arms race."

"Wow," Geo uttered, gazing at the shuttle with newfound respect.

"Something else that wasn't brought up at the presser. We think this spacecraft has the capacity to take off and land vertically like a Marine Harrier jet. We found ports in the wings that might be hiding thrusters. That might explain how no one saw it gliding low over land."

Geo scribbled that down and looked up. "Anything else?"

"One last thing. That's not dirt on the outside surfaces. It's a thin layer of dust, probably collected while it was in space. But we're not sure what it is, which is why the guard asked you not to touch. We're going to have it tested, see if it's toxic."

Geo closed his notebook and extended his hand. "Bud, I owe you. Bigtime."

5

Night was falling when he arrived at Shorty's, a busy barbecue joint on University Drive. She was sitting on a long bench on the restaurant's front porch along with several other people.

He walked up to her. "Bet you thought I wouldn't show."

She stood up and, to his surprise, gently hugged him. Her scent was musty yet enticing.

"I knew you would. I used your name for the waiting list."

"Good. I'm starved." He thought a second. "How do you know my name?"

"I just do."

He shrugged, suspecting she had seen his byline. Before they could sit down, a hostess stepped outside.

"Alejandro, you're up."

"That's us."

"My name's not Alejandro. It's Geo."

"Oh well. I was close."

Geo grinned. At least she had a sense of humor. "Right."

Inside, the restaurant was full of people gobbling down barbecued pork, chicken and ribs. Geo and the woman were seated at a laminated picnic table. While she studied her menu, Geo noticed four unsavory men were sitting at an adjacent table, gorging on their meals.

All wore work caps and sweat-stained T-shirts, apparently construction workers or day laborers. One of them, so muscular that he probably abused steroids, was staring at Geo's female companion.

Muscles, as Geo inwardly nicknamed him, nudged the man next to him. *Check that out!*

Ignoring them, Geo asked her, "What's your name?

"Gina," she answered, holding her menu with two hands.

"Got a last name?"

"Martinez."

"Where you from?"

She lowered the menu. "You writing a book?"

"Just curious, is all."

"Guadalajara, Mexico."

"My full name is George Philip Leon, but my friends call me Geo. I'm from Toledo, Ohio."

She reached a hand across the table. When he returned the shake, she squeezed and held on for a few seconds before letting go.

"Nice to meet you, Geo from Toledo."

"That's what my friends in college used to call me," he said, wondering if her handshake was a message.

Don't read anything into it. She's just being friendly.

"What you do for a living, Geo from Toledo?"

"I'm a newspaper reporter." He stopped short of telling her that he worked for a small community publication.

"That's nice," she intoned, clearly unimpressed.

A waitress came by and both ordered Bud Light in bottles to start. He wanted to ask this Gina woman why she had earlier touched him in an intimate way and invited him here. But he started with easy questions.

"What's your story?"

"My story?"

"Why did you leave Mexico and come here?"

She shrugged. "I used to run with the Mexican drug cartel, the one that guy El Chapo started." She twisted on her seat so that he could see a small scorpion tattoo on the back of her shoulder. "They branded me to show that I belonged to them."

The waitress brought their beers and they ordered dinner, Gina the barbecue chicken plate and Geo a pork sandwich. After the waitress departed, he asked, "Why in the world would you get mixed up with a rough group like that?"

She glanced at the musclebound guy, who was still staring at her. "When you grow up on the streets, you do what it takes to survive. You don't always do what's right."

That explains her tough demeanor, Geo thought. He noted that she wasn't wearing a speck of makeup. Yet, in his opinion, her face could have graced the cover of a glamour magazine. "What did you do that wasn't right?"

"You going to put this in your newspaper, Mr. Reporter?"

"Of course not."

"I transported drugs, made deliveries. Then I hooked up with this guy at the cartel who tried to pass me around to his friends like I was a bag of M&Ms or something. I decided to get out of Mexico."

The waitress brought their dinners, and Gina dug right in. As she ate, holding a chicken leg with both hands, her eyes again shifted to the muscular guy, who winked at her, while his three buddies leered at her.

"Those assholes are getting on my nerves," she said, making no attempt to lower her voice. "That big ape probably thinks I'm gonna blow him or something."

Geo feared the goon overheard that. Thanks to the restaurant's noise level, he didn't. Inserting most of a rib in his mouth, Muscles gnawed off the meat and gave Geo a don't-fuck-with-me look. Geo noticed he had a face full of stubble and acne scars.

"I imagine you have to deal with that a lot. Men staring at you," Geo quietly said to Gina.

"You wouldn't believe."

"How did you get out of Mexico?"

"I crossed the border at Lajitas, Texas, with a group of people. The Rio Grande is shallow there. The Border Patrol spotted us, but they couldn't catch everyone. So, I got away that night. Now I'm an illegal, one of the people that Americans hate so much."

Geo munched on his sandwich. "Not all Americans. How did you end up here in South Florida?"

"I heard there were lots of jobs and Latinos in Miami. So, I hitched rides with truckers to get here. If you do things for truckers, they'll take you anywhere you want."

He arched a brow. She was basically admitting to being a hooker who provided sex in return for transportation. "You did things for them?"

"I didn't screw them, if that's what you're asking. But, like I said, sometimes you do what you have to do."

Though he wanted to explore that further, he moved along. "What did you do when you got to Miami?"

"I applied for asylum at the immigration office. I heard you could do that no matter how you got into the United States. But those bastards tried to detain me. So, I had to escape. Now I got ICE on my ass," she said, referring to US Immigration and Customs Enforcement.

"What happened after that?"

"I lived on the streets for a few days. In Miami, they got soup kitchens and homeless shelters, so it wasn't bad. Then I heard about this festival at the Harley-Davidson store here in Davie. They were offering free food and beer. That sounded good, so I scraped together a few bucks and caught a bus."

Geo again noted that her English was quite good, which was inconsistent with her story about coming from the streets of Guadalajara. On the other hand, for all he knew, her mother or father was fluent in English.

"How did you scrape together a few bucks?" he asked, expecting her to say she did some "things."

"I held up a sign at a street corner that said, 'Need money for food.' I got a bunch of offers from horny men but also a lot of fives, tens and twenties."

Of course she received a lot of attention and money, Geo thought. A woman as alluring as her rarely, if ever, stands on a street corner.

"Was it worth the trip to the festival thing?"

"I suppose. They had grilled hamburgers and hot dogs. But the real reason I was there was to find work, meet someone who would give me a job."

"Did you?"

"Kind of. I told this one guy that I was living on the street and looking for work. He said he knew people who would hire me and, in the meantime, I could stay at his place. When I got to his mobile home, he got rough, tried to screw me."

"What did you do about that?"

She took a bite of chicken. "Smashed him over the head with a beer bottle. Then I stole his bike, the one I'm riding. Now this guy, a big asshole named Buzz, and his friends are hunting for me."

He liked the way she occasionally licked her lips as she chewed, something he found inexplicably sexy. "This guy and his friends are the bikers who chased you into the park today?"

"Yes. They all belong to this club. The Virile Varmints, or something."

Geo stopped eating, astounded. "You mean the Vicious Varmints. Gina, they're a notorious biker gang. They murder, rape and assault people. Their motto is 'God forgives, the Vicious Varmints don't.'"

She took a pull of beer. "Wow. I'm really scared."

"Just saying."

"Look, compared to the Mexican cartel, American bikers are pussies. I'm not worried. They'll never catch me."

Geo noticed the big man at the next table was eying Gina even more lecherously.

"Are you still trying to get a job?"

"Yes," she replied. "But no one will hire me without a green card."

"Do you have a place to stay tonight?"

Why he inquired, he wasn't sure. After the breakup with Katrina, he wasn't ready for another woman to spend the night, even if she slept in a separate room. But he didn't want to see Gina on the street, either.

"No. I bet the horny guy right there would love to have me in his bed tonight." Her eyes darted to Muscles, who was lighting up a cigarette in violation of Florida law. "But I'd prefer to stay with you."

She said this nonchalantly as though expecting Geo to offer her refuge. He wondered if that was why she had asked him to meet her here, to get free lodging.

"Any particular reason?"

"Because you're a nice guy."

"How do you know?"

"You're clean, you talk good and you got a nice face. That means you got a good heart."

"That's why you've been so friendly, touching and kissing me at the park today, hugging me and squeezing my hand tonight? You need a place to stay?"

She wiped her hands with a paper napkin. "I did all that because I want to get to know you. But if you think I'm trying to take advantage, don't worry. I can take care of myself."

He regarded her. She might have been seductive as hell. But she also was an illegal immigrant on the run from ICE and a biker gang. She was not only *in* serious trouble – she *was* serious trouble.

He knew the smart thing to do would be walk away. But she was right. He had a good heart and didn't want to see her offer herself to another man to have a roof over her head.

"You can sleep on my couch, Gina. But just for a night or two. Then you need to figure things out. Maybe check into a shelter."

She smiled and took a big bite of chicken. "I knew you were a nice guy."

After they finished eating, he paid the bill and they walked outside onto the restaurant's front porch.

"My townhouse isn't far," he said. "I'm in that Jeep over there. Follow me on your bike."

"Okay."

As they started to walk toward their respective vehicles, the muscular man, who had been staring at Gina, burst out the front door. Both Geo and Gina froze as he stomped toward them.

He got in Geo's face, so close that Geo could smell beer, ribs and tobacco on his breath.

"Get lost. Me and this babe got some business," the big thug growled. Then he grabbed Gina by the wrist and began pulling her toward his pickup truck. "We'll settle this in the backseat."

Before he could get her off the porch, Geo rammed into his back.

"She's with me, you fricking mutant!"

The collision barely budged the big hulk. With his free hand, the one not holding Gina, Muscles slapped Geo so hard that his head whipped around and he was knocked backwards.

Geo fell on his behind on the wood-plank porch, momentarily dazed. A few people seated on the bench watched without reaction. They had seen fights here before.

Gina pulled her arm free from the big man. "Shouldn't have done that."

"Yeah? Why's that?"

"Because now I'm gonna hurt you."

Muscles smirked. "That right? Now there's somethin' I'd like to –"

Suddenly he started screaming as though on fire. Putting his hands to his face, he stumbled off the porch, staggered into the parking lot and disappeared. Getting to his feet, Geo walked over to Gina.

"What did you do to that guy?"

"Maced him."

"Good work." He looked to see if she was holding a Mace canister in her hand. But she wasn't. He figured she must have stuffed it back into a pocket.

Although grateful that Gina had given Muscles a taste of his own medicine, Geo began to regret his offer for her to stay at his place. She seemed to create a ruckus everywhere she went.

Slightly taller in her boots, she gazed down at him. "Are we still going to your home?"

Her dark, pretty eyes were mesmerizing. Despite his earlier hesitation, he heard himself say, "Yes. My townhouse is on Pine Island, just north of – "

She softly kissed him and, as before, cupped a hand over his crotch. Her taste, smell and touch electrified him, even as they put on a show for the other patrons.

"You tried to save me from that creep and you're giving me a place to stay," she whispered in his ear. "I need to thank you. Let's get in your Jeep. I'll thank you here and now."

Heart pounding, he was suddenly aroused again. His biker chick fantasy was about to be fulfilled. Then he caught himself, sensing it was all wrong.

He grabbed her hands. "Why are you doing this, Gina? You don't even know me. Yet you're coming on like gangbusters."

"I know you. You're good, kind and gentle."

"Thanks, but I'm not sure I believe you."

"No? Then what do you believe?"

"I think you're using me, beyond needing a place to stay. I'm not sure for what, but this just doesn't feel right."

"Why would I use you?"

"I don't know. Maybe you're getting all lovey-dovey so that someday I'll marry you, so you can become a US citizen. If that's what's going on here, you need to find some other sucker."

Gina slowly shook her head. "You think I want to marry you after one dinner? After we met just today? You got a good imagination, pretty boy. But if that's what you're afraid of, I'm gone."

She marched to her motorcycle, pulled on her helmet and hopped on. She kick-started the powerful engine, backpedaled out of her parking spot and peeled away.

Geo watched her red taillight blend in with traffic on University Drive. He tried to tell himself that it was just as well they never hooked up. For all he knew, she was a psychopathic killer.

But deep down, he feared he had just made a horrible mistake.

6

The television networks had some fun with the mysterious space shuttle the next morning, Tuesday, the eleventh day of February. The hosts on all the morning shows implied the Russians got lost in space and errantly landed in Florida.

"Either that or they got tired of eating borscht," Michael Strahan joked on *Good Morning America*, referring to Russian red-beet soup. "The pilots, who have yet to be found, probably hit the nearest Burger King."

The cable news networks, including MSNBC, Fox News and CNN, treated the matter more seriously. They interviewed experts who maintained the spacecraft never should have entered US airspace in the first place.

"This demonstrates that our borders are vulnerable to any sort of aerial invasion," one national security expert proclaimed.

Daily newspapers and online news sites took a straightforward approach, reporting the Russians had landed on US soil with no detection or permission.

After poring over all this, Geo went to his editor's office. "Got a minute, Sam?"

"Not really," she answered without looking at him.

She was watching *The Wendy Williams Show* and nibbling on a chocolate croissant from Starbucks. Today she was clad in a pastel pink pantsuit.

He stood at the door. "Sam, I know soft features are the bread and butter of this paper. But I think we need to go after this space shuttle. It's an international story now, and the other media are all over it."

She finally gazed at him. "If the other media are already on it, why should we bother? What new information could we bring to the table?"

"As it turns out, a lot. I spoke to this FAA source who gave me a bunch of exclusive information."

"Like what?"

"This thing might be military in nature and pose a nuclear threat. It's very stealthy and apparently can land vertically. And it has this strange dust all over it. No one else has this."

Sam shrugged. "None of that sounds like that big a deal to me."

A nuclear threat isn't a big deal? Are you out of your fricking mind?

"Sam, if we do a story that includes this breaking news, it would put this paper on the map. Trust me. It's a big deal."

Sam smiled at his frustration. As she saw it, here was this former big-city reporter who thought he was hot stuff and could call the shots. But *she* was the one in control and, like it or not, Geo had to march to *her* orders.

"Forget it, Geo. I'm not interested. I think you can make better use of your time."

"Doing what?" His voice was sharper than he intended.

"There's a town council member whose hobby is model trains. He's dedicated a room of his house to them. I'd like a feature on the guy with video and photos. Today, please."

"You're kidding, right?"

"Do I look like I'm kidding? The guy's name is Roy Cassidy. He's expecting your call." She took a bite of croissant. "Now, if you'll excuse me, I'd like to finish my breakfast in peace."

Knowing that it would be futile to argue, Geo returned to his desk, fuming. His instincts told him that this mysterious shuttle involved something nefarious, and his instincts were usually right.

But instead of writing a story warning that this spacecraft might spark a nuclear war, he would be writing about a grown man who liked to play with model choo-choo trains.

It took a while, sitting at his desk, face in hands, but eventually he took a deep breath. He knew anger would get him nowhere. He would do the model trains story and make the boss happy. But first he would dig into this strange shuttle.

Screw you, Sam.

He began by researching Russian space shuttles. He found an online piece about that nation's storied space program. It contained a photo of a Russian shuttle, which looked somewhat similar to the spacecraft in the Everglades.

According to the story, the former USSR developed a space shuttle program in 1974 largely in response to the Americans, who already were building sophisticated shuttles. The Russians feared the Americans would use shuttles to construct a weapons platform in space.

In fact, the Department of Defense helped fund shuttles as a means to launch military satellites. However, the American shuttle program's primary mission was to make space flight cheaper and more routine. NASA envisioned flying frequently to a space station.

Attempting to keep pace, the USSR shot an unmanned shuttle one-hundred fifty-eight miles into space in November 1988. Called the Buran space shuttle orbiter, it flew for only an hour before landing in a desert region of Kazakhstan. It would be the only Soviet shuttle flight.

When the USSR collapsed in 1991, so did its Ministry of General Machine Building, the agency behind building a Soviet shuttle. The following year the Russians established the Federal Space Agency, which abandoned, but didn't outright cancel, a shuttle development program.

Meanwhile, after almost three decades of service, NASA shut down its shuttle program in 2011, the result of financial woes and two disasters, *Challenger* in January 1986 and *Columbia* in February 2003.

Although NASA planned to ramp up its space exploration program and fly humans back to the moon, that would be in reusable capsules, not shuttles. In the meantime, the space agency was collaborating with the private SpaceX to transport astronauts to the International Space Station.

Geo thought about all this. It had been almost three decades since the Russians reportedly had gotten out of the shuttle business. But what if they didn't? That was a lot of time to secretly construct an advanced spacecraft like the one in the Everglades.

With this in mind, he called Dave Gershwin, a senior foreign affairs officer for the State Department in Washington, DC. Gershwin was another reliable source who he had developed while working for the *Boston Globe*.

"Geo. How's it hanging, buddy?"

"Hey, Dave. It's hanging."

"I don't think we've talked since the Russians shot down that Malaysian airliner."

"Nope, we haven't. Ukrainian separatist forces took down that airliner, not the Russians."

"Who said that?"

"CNN."

Gershwin chuckled. "CNN has its head up its ass. A surface-to-air missile was used, and only the Russians possess that kind of weaponry in that part of the world. What can I do you for, sir?"

"First off, I don't work for the *Boston Globe* anymore. They downsized and I ended up at this little paper in Florida."

"What paper?"

"The *West Broward Banner.*"

Though he braced for laughter, Gershwin was sympathetic.

"At least you're employed."

"If you want to call it that. Anyway, you hear about the shuttle that landed in the Everglades a couple nights ago?"

"I don't live in a cave, Geo. What about it?"

"I hear it might be Russian and potentially a military threat, maybe even a first-strike weapon with stealth characteristics. You know anything about that?"

"I can't say anything on the record, Geo."

"Let's talk off the record, then."

"Off the record, we're trying to verify its origin. So far, the Russians are saying zilch, a big hint it's theirs. But we found no satellite evidence it was launched from Russia or any of its territories."

"Any evidence of where it was launched, then?"

"Nope. But that might be the result of those stealth characteristics you mentioned. As for its potential to be a military threat, let's just say we're already preparing counter measures."

"Counter measures? Like a retaliatory missile attack?"

"Can't confirm or deny. But you do the math."

The potential for military confrontation, triggered by the shuttle, was by itself a huge story. But it wasn't worth trying to convince Sam of that. She had already poopooed it.

"Jesus, Dave. You're basically saying this thing could start a nuclear war."

"Your words, not mine. Let's hope it doesn't go that far. Meanwhile, a team of technicians is trying to break into the ship to see if there are, in fact, any nukes on board."

"I heard."

"If you use any of this stuff, we never talked, Geo."

"As usual. Thanks much, Dave."

Geo wished he could use it. But all he could do for now was compile information and hope to eventually write a story about the shuttle and all its ramifications. To that end, he took down some notes from his conversations with Bud Kowalski and Dave Gershwin and stored them in a Word file.

Per Sam, he then called the Davie Town councilmember and arranged to see his model trains.

7

After spending an hour in Roy Cassidy's hobby room, Geo learned most everything there was to know about HO trains. Much as he hated to admit it, he found the small models to be quite interesting.

Introduced by a German manufacturer in 1922, HO – pronounced H-Oh – was the most popular scale for model trains. Serious HO operators commonly spent well over $5,000 for locomotives, cars, tracks, scenery and control units.

"It can be expensive," Cassidy said while maneuvering a model New York Central passenger train around a large table and wearing a railroad cap. "At least I'm not spending the money on cocaine or hookers. That makes the wife happy."

Geo politely laughed and wrote down that quote in his notebook, knowing it would spruce up the story. "It does look like a fun hobby."

Cassidy, a hefty man with smiling eyes, shook his head. "Electric trains aren't a hobby, my friend. They're a way of life."

Geo wrote that down too. "What does HO stand for, anyway?"

"It's kind of complicated. The 'H' stands for 'half' of what railroad modelers called the Zero scale in the early 1900s. So, it initially was the H-Zero scale, a designation for much smaller trains at the time. That eventually got converted into HO."

"Got it."

To wrap up the session, Geo took photos and video of Cassidy operating his trains and thanked him for sharing his insights.

"Good thing we have model trains because real trains, both freight and passenger, are starting to go the way of the horse and buggy," Cassidy lamented, after walking Geo to his front door.

"Just like newspapers. Have a good day, Roy."

As it was lunchtime, Geo drove to a nearby McDonald's, grabbed a couple of cheeseburgers and a large Diet Coke from the drive-thru and headed to Tree Tops Park.

Sitting at his favorite picnic table under a shady oak, he dug into the first cheeseburger and thought about Gina, the sultry Latina. He wished he hadn't offended her.

Recalling the taste of her kiss, he had no doubt that she would have slept in his bed, not on the couch, had she stayed at his home the previous night. Even though she was a sticky proposition, he wished that he had woken up next to her.

You idiot. You had to get on your high-horse. You should have had breakfast with her this morning.

Just then, he heard the roar of a motorcycle in the distance. He wished it was Gina returning to the park to find him. But no way would lightning strike twice; he wasn't that lucky, he thought, taking a big bite of cheeseburger.

Seconds later, while his mouth was still full, she charged up to him and slid to a halt, this time without a helmet, wind-blown hair falling over her face. Amazed and elated, he wanted to tell her he was sorry for being a jerk. But she didn't give him the chance.

"Get on! Quick!"

He could hear the rumble of motorcycles headed in their direction. He immediately knew the Vicious Varmints were after her again. Tossing the cheeseburger on the grass, he jumped on the bike behind her.

"Where's the police station?" she yelled as she peeled away.

Before he could answer, about twenty motorcycles thundered into view. Gina gunned it, hoping to exit the park, but wasn't fast enough. They forced her to turn into a grass field and circled her bike like Apaches trapping a settler's chuckwagon.

Gina came to a halt and extended her legs to the ground. "Shit," she spat.

"That's an understatement," Geo muttered.

A burly biker, the leader of the pack, slowly dismounted. Wearing the Nazi stormtrooper helmet, he sported a scuzzy beard and long greasy hair. His boots creaked as he walked toward them.

"Hey, Buzz. How ya doing," Gina cautiously greeted him.

"Better, now that I got your ass cornered. You stole my bike, missy."

"Stole? No, you got it wrong, Buzz. I just borrowed it."

"Bullshit. Me thinks some payback is in order. How about you start by climbing out of those jeans. When I'm done, I'll let the rest of the boys take a turn."

Despite his racing pulse, Geo got off Gina's bike and put on his best tough-guy act.

"Me thinks? You pretending to be a pirate or showing off your fifth-grade education?"

The brawny biker smirked. "You got a big mouth for a little pussy."

"Don't do it, Geo," Gina whispered.

Although he had just gotten his ass kicked by another bully – in the name of trying to be Gina's knight in shining armor – he ignored her. He approached the beefy man, who towered a good six inches above him, and put up his fists.

"Want her? You'll have to kill me first, Blackbeard."

That drew lusty laughter from the rest of the bikers, who remained on their cycles, looking forward to some good entertainment.

"Have it your way," Buzz growled.

He was about to take a swing at Geo when sirens started wailing. A parade of law enforcement vehicles stampeded into the park, lights flashing. Forming a perimeter around the bikers, dozens of officers jumped out of their cars and took positions behind their doors, semiautomatic guns drawn.

A commander got on a loudspeaker. "You with the helmet. Back away from that man," he yelled to the big hoodlum. "Otherwise, we shoot."

Buzz hesitantly took a step back, allowing Geo to exhale. The rest of the bikers froze, some hoisting their hands above their heads in surrender.

Several officers, wearing black jumpsuits, bulletproof vests and helmets, came out from behind their doors, training guns at the bikers. That's when Geo noticed they were Immigration and Customs Enforcement agents.

The ICE commander walked up to Buzz. "Get on your bike and go. We're not after you." He pointed at Gina. "We want her."

"Ain't that a coincidence, because I do too. The woman stole my chopper, officer."

"Tough shit. That's a matter for local jurisdiction. Now move along before I detain your ass for obstruction."

"Yes sir," Buzz conceded. He waved at his gang. "Move out, boys."

Some cursing, the gang members started their engines and slowly departed. Once they were gone, the commander approached Geo.

"How ya doing, George?"

"I've had better days. Hello, Bob. Good timing."

"We knew those gorillas were chasing her, so we followed them here. She's been on our radar, if you didn't know."

"I knew. She told me she attempted to apply for asylum and you guys tried to detain her instead."

Bob Klug was a senior ICE agent based in Miami. Geo had cultivated him as a source several years earlier, when Geo was with the *Boston Globe* investigating a hidden terrorist cell in South Florida. They had since worked together on several stories.

"Did she tell you she attacked two of our agents in the course of escaping from our Miami office?"

"No, she didn't."

"Came at both like a wildcat. Kicked one in the head, the other in the nuts. She's lucky we didn't shoot her on the spot."

"She is tenacious, I know that much."

"Hope you two had a nice picnic because she's coming with us now. She's facing federal charges for assault and escaping detention."

Geo looked at Gina, who had no reaction, apparently accepting her fate. Yet Geo knew she was in for a rough ride, including several months in prison followed by deportation – unless he stepped in on her behalf.

"Look, Bob. She was told she could apply for asylum without fear of being taken into custody. She had a tough time of it before coming here, being a mule for a Mexican drug cartel. She shouldn't have to face charges."

"Tell that to the judge. For now, she comes with us." Klug addressed his field agents. "Gentlemen, take this woman into custody."

Geo regarded her, the heart-shaped face, the long, shimmering hair. He knew he was about to do something absolutely insane but didn't see any other choice.

"Wait," he said to the agents, who started to move in on Gina. He turned to Klug. "She's with me now, Bob. Staying at my home. She's my … fiancée."

Klug grinned. "Right. And I'm engaged to Scarlett Johansson."

"I'm serious. I was planning to buy her a ring today. We were on our way to a jewelry store at the mall to make things official, when those goons attacked us."

Geo glanced at Gina. If she was surprised, she hid it well.

Klug wasn't buying it. "Don't piss in my ear and tell me it's raining, Geo."

"I'm not, Bob."

"Listen, you're making a huge mistake if you're trying to get her off the hook. If it's determined to be a fake marriage, you'll face federal charges and up to five years in prison. You don't want to go down with her."

Geo contemplated that but persisted. "Bob, you and I go way back. You've helped me with a bunch of stories. But I helped you too, made you look good. I'm asking for a break. Let her stay with me."

Klug studied his feet for a moment and looked up at Geo. "Tell you what. I'll back off but only if you two actually get married. I'm talking about a *real* marriage."

"How will you know it's real?"

"Seen a lot of these cases, Geo. When the preacher says, 'you may now kiss the bride,' at that point, I'll know. I'll give you a week to tie the knot. Otherwise, we'll be back for her."

"Understood. Thanks, Bob."

Klug waved his men back to their vehicles and turned to Geo. "I'm doing this only because we got history, you and me. You fuck up, you're going to find I'm a real prick."

"I already knew that, Bob. Your wife told me."

Klug grinned again and walked to his car.

After the ICE squad departed, Gina sauntered up to Geo. "Now what?" she asked.

He could tell she didn't believe he was going to buy her a ring. "Let's go to the mall. Leave your bike here."

8

Westfield mall, in central Broward County, was slow that afternoon. The big department stores were quiet and the food court was empty. Geo and Gina were the only customers at Kay Jewelers.

"So, this is what rich people wear," Gina said, surveying a case full of engagement rings.

"Or people who don't mind going into hock."

Despite overhearing those comments, a saleswoman suggested a ring with a two-carat diamond, costing more than three-thousand dollars. She placed it on a black cushion to accentuate the diamond's dazzling prism angles.

"Nice," Geo remarked.

"It's all right," Gina said, studying it.

Wishing she was more enthused Geo placed the ring on her finger to make sure it fit. "Do you like it?"

"Like I said, it's all right."

Christ. She's treating this expensive, life-changing purchase like it's a pack of gum, Geo thought. In light of everything he was doing for her, he couldn't believe she was so aloof.

Then she held up her hand to admire the ring, its juxtaposition of white gold and diamond. That's when he knew she wanted to keep it.

"We'll take it," Geo told the saleswoman.

He paid for the ring with a credit card, and Gina wore it out of the jewelry store. Seeing JC Penney was nearby, Geo realized Gina was in sore need of new clothes. She had been wearing the same jeans and halter top for days, possibly weeks, accounting for her musky scent.

"Got a suggestion," he said.

"You don't think I smell good and should get some new clothes."

It was as though she was reading his mind. "How did you know that?"

"I saw you take a sniff. I don't have money, Geo, I can't afford to buy anything."

"But I can."

Their next stop was the women's department at JC Penney. He told her to grab whatever she wanted. Although she selected only a pair of blue jeans, he suggested she get two more pairs, one black and one white.

He also helped her pick out a few blouses of different colors, fresh underwear, sandals and a nightie. Before checking out she migrated to the Sephora counter and sprayed some Chanel on her wrist, using a tester bottle. "Can I get some of this? I've never worn perfume before."

He checked the price: $105 for a small bottle. He almost said no but looked into her dark, beautiful eyes. "Why not. It's only money – that I don't have."

She kissed his cheek. "Thanks."

After throwing her new belongings in the back of the Jeep and departing the mall, they drove toward Tree Tops Park to retrieve her bike.

"Ever been married?" he asked while navigating south on Pine Island Road.

"Never let things go that far. You really going to marry me? Just to save my ass from jail?"

He glanced at her. "That's right. Just to save your ass."

"Why go out on a limb for a girl you haven't even screwed?"

"Chivalry will never die."

"Chivalry? What's that? Something you eat?"

He grinned at her guess. She was right about one thing – he was about to go out on a limb. *Way out.*

"Chivalry is where a man does something nice for a woman just because he's a nice guy."

She raised her hand again to admire the ring. "That explains a lot."

"Have you thought about what kind of wedding ceremony you want?"

"Not really. Like I said, marriage has never been high on my list."

"I'll bet you'd like a big church ceremony with lots of bridesmaids and flowers and a million family and friends."

"Right. An illegal Mexican girl, born and raised on the streets, who's run drugs and messed with badass bikers, wants a big church wedding. Get real, Geo."

"I'll tell you what I want: a small wedding with a few friends on the beach. It would look real enough to ICE because they'll have agents there."

She raised her eyebrows. "On the beach? That sounds nice."

"Then that's what we'll do."

Once in the park, she got on her bike and followed him back to his townhouse. He unlocked the door using a house key attached to a ring, and let her in.

He remained on the front porch. "I have to get back to work. Make yourself at home. There's plenty of food in the fridge."

"Thanks. I'm going to take a long, hot shower if you don't mind."

"You'll find clean towels in the linen closet upstairs."

She yet again held up her hand to study the engagement ring. "I lied before."

"How's that?"

"I love this thing."

She looked at him with heavy eyes, a not-so subtle invitation to stay home. He fumbled his key ring and dropped it on the ground.

He quickly picked it up. "Okay, then. See you later."

She smiled at his clumsiness. "Bye."

After returning to his office, it wasn't easy to write about model trains while envisioning their two bodies rocking in rhythm. But red lights again flashed in the back of his mind.

What was I thinking, offering to get hitched? And buying her a ring that will leave me in debt for the rest of my life?

He still had no idea what this woman was really about, what her ultimate motives were. Hell, if she was brash enough to steal a motorcycle from a biker, then she might be plundering all his valuables at that very moment.

On the other hand, he had acted on instinct to save her from a difficult situation with ICE. He forced himself to give his wild imagination a rest and have faith that she wasn't a crook.

He was about to start writing the model train story when his desk phone rang.

"Geo, it's Saul Finkelstein. How are you, sir?"

Geo sighed. Finkelstein was an astronomer who worked at Buehler Planetarium in Davie, giving lectures and instructing visitors how to look through a powerful telescope. He occasionally called Geo to relay what celestial events were upcoming, such as an asteroid shower or a super moon.

"Good, Saul. I'm kind of on deadline. What's up?"

"We got an interesting planetary show coming up this month."

Oh joy.

"What's it about?"

"This week Mars and Venus will be brighter than usual. Their orbits are drawing close to Earth. The following week it's Saturn and Neptune. Should make for some good stargazing."

"I'll let my editor know."

"Good, because a lot of folks are interested in what's going on in the heavens."

Bullshit.

"I'm sure."

"Speaking of which, you see all the hubbub about that space shuttle in the Everglades? I find that intriguing."

Saw it up close and personal.

"Yes, I heard about it. Pretty unusual."

"Unusual? It's freaky," Finkelstein proclaimed. "ABC and NBC are reporting that a couple of Russian cosmonauts made an emergency landing and disappeared, afraid of getting arrested."

I was told it was probably unmanned.

"Yeah, I saw the news."

"You know what I think?"

I could care less.

"What do you think, Saul?"

"I'll bet it's not Russian. Maybe it was built by some private concern here in the states or even China. But I don't think it's Russian. I saw it on television, and it looks sophisticated as hell. The Russians don't have the wherewithal to develop that kind of technology."

Yes, but they make good salad dressing.

"And here I thought they were neck and neck with us in the technology department."

"Hardly," Finkelstein countered. "Even their commercial and military aircraft lag behind ours. Here's another thing: the Russians always put a red-and-blue star on their spacecraft. This thing didn't have any markings at all."

"Maybe that's because it's top secret. Don't underestimate the Russians, Saul. The feds say they might have developed a next-generation shuttle after their initial shuttle program failed in the 1980s."

"I guess that's possible but –"

"Listen, gotta run."

"Okay, Geo. If you write a story about that shuttle, be sure to emphasize that its origin is very mysterious, no matter what the networks say."

"I would. But my editor won't let me."

"Why not?"

Because she thinks stories about petting zoos and model trains are more important.

"She has other priorities for me."

9

The townhouse was dark and quiet when Geo walked in that evening. He briefly feared that Gina had indeed purloined his possessions and fled.

But when he flicked on some lights, there she was on his living room couch, sleeping in her new black jeans and a maroon blouse. A subtle, sweet fragrance hung in the air, indicating she had applied some Chanel.

He gently touched her arm. "I'm home."

She sat up. "Hi," she mumbled, squinting at him. "I'm hungry."

"I know a good Mexican place."

She yawned. "Exactly what I wanted."

They drove to Azteca Mexican Restaurant on State Road 84 and were seated in a booth. She ordered steak fajitas and he, chicken enchiladas. Both asked for iced tea.

After a waitress delivered their drinks, he asked, "What did you do with your old clothing?"

"Threw it in the trash."

"Good idea." He drank some tea. "Tell me about your life in Guadalajara."

"You still writing that book?"

"It would probably be a best seller."

She poured two packets of sugar into her iced tea and took a sip. "Let's put it this way. I won't go back."

"Why not?"

"I always felt threatened. Men were always trying to corner me, even when I was a young girl. That's why I ended up with the cartel. They at least provided some protection."

"What about your parents? They didn't protect you?"

"My mother worked in the sugarcane fields and my father died before I was old enough to walk."

"Any brothers or sisters?"

"Two sisters and two brothers, all older. I haven't talked to any of them in a long time. None of them amounted to any good."

He got the picture. She had led a difficult life.

"May I ask, how did you learn to speak English so well?"

"I was a good student in school until I dropped out. Let's talk about you."

He found it hard to believe that a young Mexican girl struggling to survive on the streets could become fluent in another language, particularly with a limited education. She was obviously hiding something about her past.

"What do you want to know?"

The waitress delivered their dinners. Gina built a fajita, piling sliced steak, grilled onions, green peppers, grated cheese and guacamole onto a soft, warm tortilla. Holding it in two hands, she took a big bite. "Damn, this is good," she moaned.

Geo smiled, enjoying the provocative way she ate. He picked up his knife and fork, cut a small piece of enchilada and carefully inserted it into his mouth.

She finally answered his question. "Everything. That's what I want to know about you. Pass the *picante* please."

"The what?"

"The hot sauce."

He did so, and she shook the red sauce all over her fajita. He expected her to spit out her next bite, her mouth on fire. But she closed her eyes in pleasure.

"This is the best," she declared.

"Isn't that kind of spicy?"

"It's delicious. You Americans can't take the heat."

"At least this American can't."

"Are you going to tell me about yourself, or what?"

Geo sighed. "Not that much to tell. I grew up in Toledo, went to an all-boys prep school, got my bachelor's degree from the University of Florida and a master's in journalism from Northwestern."

"Northwestern. Where is that? In the Northwest?"

Geo grinned at her geographical guess. "Chicago. Then I worked for the *Boston Globe*, got laid off and did a stint with a supermarket tabloid. To get back into respectable journalism, I applied to the *West Broward Banner*, and, well, here I am."

"You been all over the place. Toledo, Chicago, Boston, Florida."

"That's me. A travelin' man."

"Maybe someday I'll travel too, take a long trip."

"I wouldn't exactly call your journey from Mexico a puddle hop."

"Whatever. Got any family?"

"My mom lives in a senior community in Pompano Beach. My older brother, Mickey, is a car mechanic in Los Angeles. That is, when he's not gambling or drinking. Or in jail."

"What about your father?"

"He was a combat pilot in Vietnam, flew the F-4 Phantom. After the war, he became a traveling salesman and died in a DUI traffic accident."

"Tough break."

"I'm over it. It was several years ago."

"Got a girlfriend?"

"How could I? I'm about to marry you."

"Got a girlfriend, or not, pretty boy?"

"I did. But she's long gone," he said, lying, as Katrina had just dumped him.

"What do you do for fun?"

Geo shrugged. "Truth is, I don't have that much fun. I swim or walk for exercise. Sometimes I drive my Jeep up US 27 to get away from the chaos. Otherwise, I mostly work. What do you do?"

"I'm here with you. That's all the fun I need. Why do you drive on US 27?"

"It's a major highway that sees little traffic and aims into open territory. I like open spaces."

"Sounds like you'd enjoy flying to the moon."

"If I could make a reservation, I would."

That made her giggle. He hadn't heard her giggle before. It was a sweet, gentle sound, which seemed odd for such a tough woman.

"I like open spaces, myself. Make reservations for two."

"Deal."

After they finished eating, out of the blue, Gina said, "I didn't show up at the park today because I needed help with those bikers. I wanted to see you."

He remembered how she had left in a huff the previous night. "Why?"

"Because I knew you didn't want to say goodbye."

"No, I didn't," Geo quietly confessed.

He suddenly wished they had met under normal circumstances, like at a bar or coffee shop, and taken time to develop a real relationship. But here he was, about to walk into a contrived marriage to save her from the law.

She held up her left hand and gazed at her engagement ring. "I don't want to take this off, not after what you did for me today."

Geo smiled. Despite everything, that was good to hear.

After paying the check, Geo followed Gina toward the restaurant exit. She abruptly stopped and kissed him on the lips in front of tables full of people.

"I really enjoyed dinner," she said. "Now let's do something fun."

"Fun? Like what?"

"We'll think of something."

He remembered that she wanted to "thank" him for saving her from ICE. He wasn't sure he was ready for that. Though attracted to her, he didn't want to get entangled in another emotionally bogus relationship.

As they walked toward his Jeep, he was thinking of excuses to avoid fooling around once they got home. Gina pointed at a building across the street.

"That looks like a dance club. Let's check it out."

"Gina, I'm not much of a dancer."

She grabbed his hand. "Let's go, pretty boy."

"Okay."

The club was called Casa Salsa. Through its storefront window, they saw about a dozen couples dancing to fast-paced Latin music provided by a live band.

"I'm telling you, Gina. I look like a spastic monkey when I dance."

"Too bad."

After Geo paid a cover charge, they joined other couples on the dance floor. She began rocking her hips to the salsa music. He tried to imitate her rhythmic movements but felt clumsy and out of place.

"You're doing fine," she called out to him. "You don't have to be Mexican to do this."

"I'm not sure about that."

Eventually he got the hang of it, his footwork following hers. But after a few dances, sweat was beading on his forehead. "Gina, why don't we take a break."

"Okay, monkey boy. Let's sit the next one out."

"Thank God."

Just then, the lights dimmed and the band began playing a soft, melodic tune. Couples drew close to dance slow. Gina wrapped her arms around Geo's neck, rested her chin on his shoulder and held him tight. His heart started pounding.

When the song ended, she said, "Let's go home."

"Home?"

"Home. You know. Where you live. Where we can be together."

Does she really want to do this? Do I?

"Okay," he relented.

During the drive to his townhouse, his jitters and excitement were growing as she calmly held his hand. Once inside, she charged upstairs to his bedroom.

"Have to go to the bathroom," she yelled. "Those fajitas got to me. Don't come up."

"Right," he said, at the same time relieved and disappointed. If she had the runs, no way would she be in the mood for romance. Probably for the best, he tried to convince himself.

Twenty minutes later he was sitting in his recliner in his den, watching a college basketball game, when she came downstairs. She was wearing her new nightie. Her hair was loose around her face. Her eyes were inviting.

He swallowed as she walked over to him. Hiking up her nightie, she straddled him. He saw that she was wearing the new panties he had bought for her.

"It's time to thank you."

She kissed him gently. Once again, that electric thrill filled his entire body.

"Gina," he rasped, "I want you bad. But I want to make sure you want me too, that you're not doing this just to thank me."

"Trust me," she whispered. "I want you."

10

Geo arrived at his desk early the next morning to get a jump on the day, only to find Sam had fed him another goofy assignment.

"The quarterback at Cooper City High has taken a job at McDonald's on Griffin Road," she wrote in an email. "He's an All-American being recruited by several universities. Pretty cool that a future NFL star is working at a fast-food joint. I'd like a feature, photos and video by end of business today, please."

Cool my ass. He probably just wants free Big Macs.

Geo didn't fret for long. He was still savoring the previous night. Although he expected Gina to get hot and heavy, they had only kissed and cuddled. Even so, her passion had been breathtaking, particularly when she –

His desk phone rang, busting him out of his reverie. It was Bud Kowalski, the FAA field agent.

"Morning, Geo. We're going to try to enter the spacecraft today, but it's going to be tricky."

"Why is that, again?"

"As I told you, we suspect the shuttle is military in nature and might contain a nuclear warhead. If so, we sure as hell don't want to set it off."

"If there's a nuke on board, wouldn't you be able to detect it from the radiation?"

"Not necessarily. The outer skin appears to be so impermeable that it would prevent radiation from seeping out. We're hoping we can finagle open a hatch with drills and power saws."

"What happens if you get inside?"

"We'll be able to see what this thing is all about, hopefully confirm it's Russian, in which case they'll have a lot of explaining to do. Want to watch?"

"Sure. But what if you guys detonate the nukes? Wouldn't we all get roasted?"

Kowalski released the raspy chortle of a smoker. "If we detonate the nukes, all of South Florida will be obliterated and probably Cuba too. We're holding a presser this afternoon, but no media will be allowed at the airport this morning. Only you."

Geo thought about the potential for a huge, exclusive story. "I'm on the way."

He was about to jump in his Jeep when Sam rolled up in a blue Jaguar SUV, a $60,000-plus vehicle. Sporting designer sunglasses, she parked next to his vehicle, checked her makeup in the rearview mirror and turned off the engine.

Although he knew Sam was paid a hefty salary, Geo guessed that publisher Joe Cronk had helped finance the Jag.

Wearing a white skirt suit, she got out of the SUV holding a latte purchased at a Starbuck's. "Where you going?"

He didn't lie. "To that Everglades airport. The feds are trying to crack open that shuttle to see if any nukes are on board. That's a big story, Sam, and we're exclusive until the feds hold a presser later today. We can get something online before the press conference starts."

Sam closed her door, locked the vehicle with a fob and walked up to him, close enough that the aromas of her sweet perfume and coffee filled his nostrils.

"You're pushing my limits, Geo. I told you, we don't do hard news."

"Yes, I know. But this warrants an exception. Look, Sam. I'll get the Cooper City quarterback story done by this afternoon. Promise. But I'm asking nicely: can I check this thing out? It's important."

She breathed in deeply. "Don't stay out there long. I want the quarterback story as early as possible. After it's written, if you have time, you can post something about that stupid shuttle."

Right, that stupid shuttle, which is potentially carrying a nuke that could blow much of Florida off the map.

"Deal. Thank you, Sam."

She mumbled something unintelligible and walked inside.

When he arrived at the Dade-Collier airport, military guards stopped him at the entrance, asked to see his credentials and directed him to park near the command center tent.

Hopping out, he leaned against his vehicle and watched as technicians attempted to open a hatch behind the shuttle's cockpit with various tools. A few minutes later Bud Kowalski emerged from the tent and walked over to him.

"Fricking Russkies. What a major league pain in the ass they are," Kowalski grumbled, a cigarette dangling from his lips.

"Morning, Bud. What's the problem?"

Kowalski took a drag. "Nothing's working. Not drill bits, power saws or even a blowtorch. The skin is like a cross between high-grade steel and an advanced polymer. Whatever it is, it's impenetrable."

Geo pulled a notebook and pen out of his jacket pocket. Although he had a digital voice recorder, he preferred the old-school method of jotting down notes.

"Is there any kind of lock, where you insert a key?"

"This thing is far too advanced for that. We've searched the fuselage for a keypad that would open the hatch with a code. But no dice. The Russkies really outdid themselves this time."

"What now?"

Kowalski blew out a plume of smoke. "Not sure. Everything about this spacecraft is incredibly sophisticated. The skin, engines, wings and windows all incorporate designs and materials we've never seen before."

"Damn."

"The wheels alone are locked so tight that we can't even tow the damn thing. That's not to mention they're made of super-hard fabric, not rubber. In some areas of technology, the Russians are kicking our asses."

"You've confirmed it's Russian, then?"

"We haven't confirmed anything. But we can't find another explanation. No other nation on Earth has a space program that could produce this."

Geo remembered his conversation with Saul Finkelstein, who didn't think the Russians were involved. "I got an astronomer source who says a private company or China likely developed it."

Kowalski shook his head. "Sure as hell wasn't the Chinese. They're focused on putting people on the moon. And if a private concern had developed something this advanced, they would have trumpeted their achievement to the world by now."

"What about a maverick like Jeff Bezos? Maybe he built it."

"Who's Jeff Bezos?"

"The guy who started Amazon and has so much money that he had a contraption built to carry a crew to the moon. The Blue Moon lander, I think he called it."

Kowalski rolled his eyes. "This thing is not American in any manner, shape or form. Of that, I am quite certain."

"Looks like the Russians, then."

"Yep. The fricking Russkies."

Geo closed his notebook and clicked his pen. "Any results back from that dust?"

Kowalski dropped his cigarette to the ground and snuffed it out. "Not yet. But I can tell you right now, there's something weird about that stuff."

"How so?"

"It has a slippery feel, like oil residue. It's unlike any dust I've ever seen before."

"What could cause oil residue in outer space?"

"I'm not sure. But it's probably not oil. It's something more akin to mercury. As in the compound, not the planet."

"Mercury? Isn't that the stuff they put in thermometers?"

"Yes. But like other elements, mercury can exist as a liquid, solid or gas. For instance, at forty degrees below zero, mercury freezes into a solid. Meanwhile, mercury vapor can be found in particles of light. I'm guessing this is vapor."

"Interesting. Gotta run. Thanks again for everything, Bud."

"Try not to make us look too stupid in your story."

"Wouldn't think of it."

As he drove back toward civilization, Geo called the Cooper City quarterback and arranged to meet him at the McDonald's where he had been hired. Geo thought it was a dumb story because thousands of students worked at fast-food joints.

Yet, after arriving at the restaurant, Geo found this kid, whose name was Dean Baker, had a compelling story.

"I was a little runt in elementary school," Dean told Geo, while the two sat at a corner table. "One day, this bully beat the snot out of me in front of a bunch of kids. They all laughed. That's when I decided to bulk up and play football."

"Did you ever pay back that kid for what did?"

Dean smiled. "No. He's the fullback on our team now, and if it weren't for him, I wouldn't be where I am. So, when he apologized, I forgave him."

A few giggling high school girls moseyed over and interrupted the interview, flirting with the tall, young quarterback.

"What a pain. They're always hanging out here," Dean said to Geo after the girls left.

Because you're here. The restaurant manager probably hired you to drum up business.

"Where do you think you'll end up playing? Ohio State? Alabama?"

Dean shook his head. "I want to play for Miami. They got it going on with their young coach and that turnover chain."

Geo nodded. The story just got better yet. The kid wanted to remain in South Florida, bolstering the local angle. "Good choice. Do you eat here for free? Is that a perk?"

"We get one free meal per day. But that means I'm only allowed to have one Big Mac and a large order of fries per day."

"That's a lot food."

"For you maybe. I could murder about five Big Macs and five orders of fries, no problem."

Geo politely smiled, making note that in young people's parlance, "murder" meant "devour."

Geo shot photos and video, wished Dean luck in his football career and headed to the office. After settling at his desk, he called Dean's coach and parents and even the bully for color. He then batted out the quarterback feature to appease Sam.

Taking no break, he next composed a story saying federal officials had all but confirmed the space shuttle was Russian because of its advanced technology. The story noted that authorities were unable to breach the ship and feared that it might be military in nature, potentially a first-strike weapon with a nuclear warhead.

The staff's online producer posted both stories on the *Banner*'s homepage. While the Cooper City quarterback was given prominent play, the shuttle was minimized at Sam's direction. She further removed the part about the shuttle potentially carrying nukes.

Sam sent Geo an email, saying including such volatile information without absolute confirmation was "unprofessional and irresponsible."

Seeing all this, Geo was so pissed off that he almost wrote a retaliatory email, saying only an idiot would give a soft feature about a high school kid higher priority than an exclusive about the shuttle posing a threat to national security.

Instead, he drove home, likely saving his job in the process. At least he had Gina waiting for him.

11

He was turning into his townhome complex when Katrina Tucker called. He briefly debated whether to answer. Hadn't they already said goodbye? He answered anyway.

"Hey Katrina. What's up?"

"Come over tonight, Geo."

Late afternoon sun mingled with high, feathery clouds. He parked the Jeep and shut down its engine. "Why?"

"I want to be with you. One more time. Maybe more."

Aren't you the woman who told me how black you are – and how white I am?

"Look, Katrina. You broke it off, remember? I don't understand why you're even calling."

"I was thinking about you, what it's like to be close to you. I don't want to give that up. Call it friends with benefits if you want."

He jumped out and leaned against his vehicle. "I don't want to be your friend, Katrina. I wanted to be your soulmate until you made it clear that's not on the table."

"I told you I can't marry you, and I told you why. That doesn't mean we can't make each other happy, at least until I get married."

He released a cynical chuckle. Although eager to be intimate with him, she wanted to marry another man, someone who she probably had yet to meet. It was obvious that she considered Geo a boy toy and always had.

"Do you love me, Katrina?"

"Love doesn't have anything to do with it, Geo. I have to think about what's best for my future, for the future of my children."

"I'll take that as a no. If you really loved me, the black-white thing wouldn't matter. Again, goodbye, Katrina."

He ended the call feeling both victorious and deflated. As someone once said in a song, breaking up is hard to do. He now knew that to be a fact.

Once inside his townhome, he was about to grab a beer, take the edge off, when he saw her lying on the living room couch, gazing at him. She didn't have to say anything. Her body language said it all.

I want you. Now.

Perhaps it was because Katrina had just reopened a fresh wound. Perhaps it was all the frustrations surrounding Sam. But he wasn't in the mood for intimacy, even though Gina looked sumptuous.

"Let's go bowling."

She sat up. "Bowling? What the hell is that?"

"They don't have bowling in Mexico? It's fun. You roll a ball down an alley and knock over a bunch of pins. And the bowling alley serves great pizza. We can have dinner there."

"Pizza?"

"We'll make a night of it."

She sprang to her feet. "Okay."

Bowlero Davie, on State Road 84 near University Drive, was bustling with families, kids and couples. The lighting was dim, accentuating a series of entertainment screens. The delightful aromas of grilled hamburgers, French fries and pizza filled the air.

Gina and Geo were assigned a lane near the concession area. He rented bowling shoes, selected a couple of balls and set up their electronic scoreboard. "You're up," he said, sitting at a table behind their alley.

She observed other people rolling balls down their alleys, the crash of pins resonating throughout the cavernous building.

She almost seemed frightened. "I've never done this. What do I do?"

He smiled, finding it both funny and heartwarming that a tough biker chick was intimidated by something as convivial as a bowling alley. "Let me show you."

Grabbing a ball from the return machine, he demonstrated how to put fingers in the holes. He lined up, took a few steps, swung his arm back and released. The ball bolted down the alley and collided with the pins. Strike.

He pumped a fist and walked back to her. "That's how it's done." He fetched a light ball for her. "Your turn."

"Show me how you move your arm," she said, placing her fingers in the ball.

He had her face the pins and got behind her. He took her arm and gently brought it back, careful to maintain some space between their bodies. This was a family place, after all.

"Keep your arm straight as you follow through," he said, backing away. "Give it a shot."

She took three steps forward, swung back her arm and jerked it forward. She heaved the ball three-quarters down the alley before it banged down. Heads turned in their direction as the ball smashed into the pins, knocking them all down.

"Strike!" she blurted, pumping her fist and dancing around in a circle. "That wasn't so hard."

Geo held up his hands to the employees at the desk, who were watching intently. *Don't worry. We'll tone things down.*

"That was excellent," he said as she ran to hug him. "But next time, try to put the ball right on the alley rather than throwing it. Okay?"

"If you say so."

Thereafter, she bowled with more constraint but still released the ball with astonishing accuracy, at least for a rookie. They played four games and she beat Geo in the last one by more than twenty points.

"Time for pizza," he mumbled.

He ordered a large pepperoni pizza and two bottles of Heineken at the concession counter. As they drank the brews, waiting for their dinner, he asked, "You sure you've never bowled before?"

"I never even heard of this game before tonight."

"Guadalajara doesn't have bowling alleys?"

"Not on my side of town."

When the pizza was ready, they took it to a table. He put a slice on a paper plate for her. "Careful. It's really hot."

"I like hot." She took a big bite into the triangle. "Oh my god. This is so good."

He again loved the way she relished food. "It's too bad that something as tasty as pizza isn't very healthy."

She took another bite. "Who cares about health."

When they finished, she wiped the sauce off her mouth with a paper napkin. "Let's go home."

His stomach tied into a knot. Although they already had been close, at least to a point, he was hesitant to consummate their relationship. Why, he wasn't sure. Possibly because he feared that doing so would make their union legal once they were married.

Then he studied her, saw how ravishing she looked in the subdued light, and realized he was kidding himself. He wasn't concerned about the legal ramifications. He was afraid of falling head over heels in love and careening into an uncertain future.

It's time to take a chance because you may never meet another woman like this again.

He nodded. "Yeah. Let's go home."

During the ride back his townhome, she held his hand and nonchalantly looked out her window. He was having a hard time swallowing because of the lump in his throat.

The moment they were inside his unit, she closed the door and kissed him hungrily in the front foyer. She stopped and looked into his eyes.

"I'm grateful for what you've done for me, buying me a ring," she whispered. "But I'm doing this because I want to. Believe me?"

"Yes."

Taking his hand, she led him to the living room couch, where she first undressed him then shucked out of her own clothing. In cozy dimness, the only light coming from the kitchen, she pushed him down on the couch and straddled him. Their two bodies became one, and it was like someone plugged his entire body into an electrical socket.

"What about birth control?" he rasped.

"I've taken care of that," she assured.

She gently gyrated her hips, letting him savor the warmth and feel of her. Then she escalated the rhythm and intensity, sending him floating to Mars. As she pounded up and down, she raised her chin toward the ceiling. He wrapped his arms around her back and held on tight.

At the moment of release, he bellowed in ecstasy, and she howled at the top of her lungs. She collapsed on top of him and buried her face in his neck, both puffing.

A satisfied smile spread across his face. He had been with a lot of sexy, attractive women. But this had been the most erotic ride of his life. He closed his eyes, about to drift off, when she whispered in his ear.

"Let's go again."

He awoke a few hours later in darkness, save for moonlight filtering through window blinds. Sitting up he saw Gina lying next to him. They had moved from the couch to his bed for their second act.

She was lying on her stomach, hair spilling over her pillow, partially covered by a blanket. He studied her lovely face, the ample curve of hip and the long stretch of thigh. She was dazzling even as she slept.

Suddenly he no longer cared that she led a dangerous life, hunted by bikers, the Mexican cartel and ICE. He was too infatuated to walk away. All he could do now was protect her as best he could and deal with whatever trouble lay ahead.

Unable to sleep, he turned on the television and kept the volume low. An old black and white movie was just starting. According to his cable provider's information, it was entitled *It Came from Outer Space,* released in 1953. Having a taste for old movies, he settled back and watched.

It opened with a stargazer named John and his girlfriend observing a flash of light streak across the sky. They chase after it and discover an alien spaceship has crashed in the Arizona desert.

The aliens turn out to be single-eyed, jellyfish-like creatures. They start abducting people in a small town and replacing them with extraterrestrial duplicates. But John learns the aliens aren't hostile.

Rather, they pretend to be the humans who they have abducted in order to collect materials to repair their spaceship. John protects the aliens from a trigger-happy sheriff, allowing them to fix their ship. In return, the aliens return the captives.

After the aliens depart, John's girlfriend asks if they are gone for good. "No, just for now," he responds. "It wasn't the right time for us to meet. But there will be other nights, other stars for us to watch. They'll be back."

As the movie ended, Geo smiled, wondering how its writer came up with such a bizarre plot. Turning off the television, he wrapped himself around Gina.

"That movie was so phony," she murmured, half asleep. "Space aliens don't look like that."

"No? What do they look like?"

"Little green men, of course."

12

Geo was sitting at his desk on Thursday morning, munching on a jelly donut and waiting for Sam to dump another stupid assignment on him. Having gotten little sleep the night before, he was exhausted yet exhilarated.

Gina had awoken him before dawn and again sent him flying straight into the stratosphere. Closing his eyes, he relived the thrill of how she had –

His desk phone started ringing. He pulled on his headset and answered the call.

"Morning, Geo. It's Bud Kowalski. We got the results back from that dust."

"The dust?"

"The dust covering the space shuttle."

"Right. That dust. What'd you find?"

"It's from deep outer space. And I mean deep."

Geo put down his donut and picked up a pen. "Meaning what? The thing was flown in by extraterrestrials?"

Kowalski laughed. "Not *that* far. But it is highly unusual. It indicates the Russians have really been out there."

"Like how far?"

"At least as far as Jupiter, or five-hundred sixty million miles from Earth. It would take a spacecraft about three and a half years to travel that distance. But it might have gone even farther."

"That thing went all the way to Jupiter with Russians at the controls?"

"At this point, there's no evidence that anyone was on board. On the other hand, quite frankly, we just don't know how it was controlled."

"So, the thing zips out to Jupiter and ends up in the Everglades. Weird."

"Again, it probably made an emergency landing due to some kind of mechanical problem."

Geo thought a moment. "If it did go into deep space, doesn't that reduce the likelihood of the shuttle carrying nukes for military purposes?"

"Not necessarily. It's possible the Russians conducted tests in space, where nukes would cause no harm. They also probably wanted to test the ship's long-range capabilities."

"You tell any other media about this?"

"For now, only you. But if the *New York Times* calls, I won't hold back. And they call every day. So, you'll be exclusive only for so long."

"Got it. What happens now?"

"We keep searching for how and why this thing is here. We keep trying to verify the Russians are behind it. We keep trying to determine if it's carrying an advanced weapon or is a weapon by itself."

"And if it is?"

"Then we have an international crisis on our hands."

Geo thanked the FAA field agent for the call, walked over to Sam's office and sat down. He didn't bother to say good morning.

"You know that shuttle in the Everglades? It might have gone as far as Jupiter. So I was just told. That's not exactly a trip to the grocery store, Sam."

She shrugged. "Big deal."

Big deal? The woman has the news judgement of a ripe avocado.

"Look, Sam. This time, we absolutely need to do a story. The feds say a coat of dust indicates it's been into deep outer space, where it might have conducted military tests. Odds are growing this thing is a first-strike weapon, and for now, we're exclusive."

Until he interrupted her, she had been leafing through *Cosmopolitan*, searching for ideas on how to fortify her wardrobe. She planned to hit the mall during her lunch break.

Now she was agitated. Geo was wasting her time again. She wasn't about to let him write another shuttle story, not after the first one spurred Joe Cronk to send her an angry email.

"Why was Geo allowed to do a piece that counters our mission to produce only highly local news?" he had written. "He's hurting us, Samantha."

Sam put down her magazine. Her tone was cool. "I don't care whether we're exclusive or not. I don't know how many times I have to tell you this – people buy our paper because we leave murder and mayhem to the big boys."

"Sam, we'd be out of our minds to ignore this thing; it's too important."

"Then we're out of our minds. Forget about the shuttle. I need you to run over to the Aldi grocery store. They have a seven-pound tomato for sale, possibly the biggest one in Broward County history. I want a feature on whoever buys it and how they plan to use it."

"A giant tomato," Geo muttered. "You want a story on a giant tomato rather than how this mysterious space shuttle potentially flew as far as Jupiter."

"That's exactly what I want."

"Are you out of your goddamn mind?" he yelled. "We have the opportunity to do a story with global implications and you're tossing it aside in favor of a stupid tomato?"

The other reporters looked up, hearing Geo's outburst. They smirked, enjoying the show in Sam's office.

Sam remained calm. "Geo, let me give it to you straight. If you want to keep working here, you need to stop disputing me. Either play by my rules, or I'll find someone who will."

That hung in the air like a bad odor. She had never threatened to fire him before. He had always assumed his years of experience, particularly in comparison to everyone else on the staff, made him too valuable to let go. Now she was telling him in no uncertain terms that wasn't the case.

Undaunted, he stood up and put his palms flat on her desk. "Let me give it to *you* straight, Sam. I'm working at this little paper because I need the job. I do whatever you ask because I'm a professional. But I warn you, without more latitude, I'm out of here. I'd rather flip burgers at McDonald's."

Not expecting that response, Sam almost snarled, *Don't let the door hit you in the ass.* But the last thing she wanted to do was lose George Leon.

She motioned for him to sit down, which he did, hesitantly. She knew that what she was about to say wouldn't sit well with Joe Cronk. But she would deal with the publisher later.

"Tell you what, Geo. You can write about that shuttle. But only after you've completed my assignment. Be forewarned, I am *not* guaranteeing how I'll play your shuttle story. We clear on that?"

He figured this concession was a major victory, given Sam's normally unbending nature. "Fair enough. I'll do the tomato story this morning. Then I'll work on the shuttle story."

"Fine."

As Geo walked to his desk, Sam exhaled, inwardly patting herself on the back for not letting things get out of hand. She picked up her *Cosmopolitan* and continued leafing through it.

Geo, meanwhile, fell down in his chair and stared out his window. A horseback rider was waiting in line at the McDonald's drive-thru. Moe's Garage had two pickups elevated on lifts. A fender-bender was causing a traffic jam on Davie Road. He didn't see any of it.

Although Sam had relented, he was still deeply frustrated. He wished he was working for a real newspaper where he would be encouraged to go after a big story like the shuttle. But here he was, stuck in a small shop.

Then he thought about the previous night with Gina, how good it was, how he couldn't get enough of her. He recalled the crazy movie about creatures from outer space and Gina saying all aliens look like "little green men." That made him smile.

Being curious, one of the traits that made him a top investigative reporter at the *Boston Globe*, he decided to find out where the term "little green men" originated. With some time to spare, he checked an online site that specialized in extraterrestrials.

It explained that "little green men" became popular in the 1950s, when people began claiming they had seen flying saucers, which was all the rage back then. In one instance, in 1955, two men reported an encounter with four-foot-tall, metallic-silver aliens in rural Kentucky. Having a little fun, newspapers altered their story to say the aliens were "little green men."

Chuckling, Geo mentally braced to transition from little green men to a big red tomato. Yet he wondered what extraterrestrials, if they in fact existed, looked like. Did they resemble the horrible creatures portrayed in Hollywood movies?

Shaking his head, he realized it was a moot point. He didn't believe for a second that extraterrestrials from another planet would ever visit Earth.

13

As was the case with the model trains, the giant tomato story was more interesting than he expected. The tomato in question was only a pound shy of the largest tomato in US history. That behemoth, grown in Walla Walla, Washington, in 2018, weighed a whopping 9.4 pounds.

Like the one in Walla Walla, the Broward County tomato was so big that it resembled a small, red pumpkin. An elderly man, who was born in Sicily, Italy, and loved to cook, bought it from the local Aldi's grocery store.

Geo asked him what he planned to do with it. "I could make one hell of a lot of BLTs," the man answered with a wry smile. "But it will make a nice spaghetti sauce for the family dinner. All I need is a couple bottles of Chianti."

Upon returning to the office, Geo whipped out a tasty feature about the giant tomato. Then he focused on the shuttle. Because it had rocketed deep into space, as evidenced by the strange dust, he wanted perspective on a trip of that magnitude. He called Saul Finkelstein, his astronomy source.

"Tell me, Saul, what's involved in flying a spacecraft to Jupiter? I ask because that's where the shuttle in the Everglades has been, apparently."

"They think that shuttle has been as far as Jupiter?"

"Maybe farther."

"With humans on board?"

"That's unknown. They still haven't found any astronauts or Russian cosmonauts. But the feds tell me there's a slim possibility that people were flying it."

Finkelstein laughed. "Listen, it's not unusual to send a spacecraft as far as Jupiter. Pioneer 10 and Pioneer 11 flew near Jupiter in the 1970s.

Voyager 1 flew beyond our solar system in 2012. But if you're talking about with people on board, that's inconceivable."

"Why?"

"It would take about thirteen-hundred days to fly to Jupiter, or three and a half years, and another thirteen-hundred to return. You're taking about a seven-year mission not including the time spent exploring the planet."

"Yikes."

"The logistics of providing enough food, water, oxygen and fuel for that duration would require a spacecraft the size of a cruise ship. That's not to mention Jupiter is the last planet you'd want to visit. It's extremely treacherous."

"Why is that?"

"Can I give you some background?"

"Sure." *As long as it doesn't take too long.*

"Jupiter is the fifth planet from the sun, named after the king of the ancient Roman gods," Finkelstein commenced, sounding like he was giving a lecture at the planetarium. "Yellow-orange in color, it's the fourth brightest object in the night sky behind the moon, Venus and Mars."

"That's interesting." *Not really.*

"Jupiter is larger than all the other planets combined and has an enormous orbit around the sun. One Jupiter year is equal to almost twelve Earth years. Yet one Jupiter day is equal to only ten Earth hours because it spins almost twice as fast as Earth."

"Fascinating." *Get to the point, dammit.*

"Here's why it's treacherous. Jupiter is composed mostly of hydrogen and helium. Whether it has a solid core is unknown. But it definitely doesn't have a solid surface. So, if a ship flew into its atmosphere, extreme temperatures and pressures would vaporize it."

"Vaporize it?"

"Poof. Just like that."

"That does sound a bit foreboding."

"Here's what's really weird about Jupiter. It has seventy-nine known moons, four of which are considerably larger than the rest. One of those four may be hiding an ocean of warm water beneath an icy crust."

"Why is that weird?"

"Because even though Jupiter itself is incredibly cold and hostile, that one moon, called Europa, might be warm and wet enough to sustain life."

Geo contemplated that. "Is it possible the Russians wanted to explore that moon? Maybe as their way of pulling ahead of the Americans in the space race?"

Finkelstein laughed again. "There is no space race, hasn't been for decades. And while that single moon might not be as perilous as Jupiter, it would still be horribly risky to travel there, particularly when it would be more productive to explore Mars."

"Oh."

"Let me tell you what's a lot more realistic than either the Americans or the Russians going to Jupiter."

"What?"

"That shuttle could very well be an alien ship from outside our solar system, that's what. That's the real story here, my friend."

Geo cracked a grin. "Maybe it was flown here by the same aliens who tried to blow up the Earth in *Independence Day*," he said, referring to the 1996 movie starring Will Smith and Jeff Goldblum.

"I'm serious, Geo."

"Christ, Saul. This isn't a movie or a video game. Aliens are about as real as the tooth fairy. ET sure as hell didn't land that shuttle in the Everglades. Humans did."

"All evidence to the contrary, Geo. Humans lack the technical know-how to construct something as cutting-edge as the ship in the Everglades. As far as I can tell, it's far more sophisticated than anything that's *ever been built* on Earth."

"What do you know about its construction?"

"I saw photos of it. I also have a friend at NASA who tells me its exterior is so tough the feds were unable to get in. Plus, they can't find portals for the engines that allowed it to make a vertical landing. Yet that's just what it did. All of this indicates an advanced alien race flew that thing here."

Geo rolled his chair back from his desk to stretch his legs. He was having a hard time taking Finkelstein seriously. "My source tells me the Russians might have developed that kind of technology."

"Bullshit. The Russians can't even put a man on the moon. You know, a lot of astronomers, myself included, think we'll make contact with extraterrestrials by 2035. Looks like it happened ahead of schedule."

Geo knew that Finkelstein was an expert in all things celestial. He had graduated from the Massachusetts Institute for Technology with degrees in rocket science and astronomy. He then worked at NASA as a space engineer for several years.

Yet it was no secret that Finkelstein was slightly off his rocker. NASA fired him after he attempted to stow away on board a shuttle. Subsequently, the only job he could find was as a glorified tour guide at Buehler Planetarium.

Nonetheless, for fun, Geo pressed Finkelstein about the possibility of a space alien visit. He pulled his chair close to his desk again.

"Let's say that extraterrestrials actually flew the shuttle here. Where do you think they came from?"

"Possibly TRAPPIST-1. That's all caps, thanks to the weird nomenclature of modern-day astronomers."

"What the hell is TRAPPIST-1, all caps?"

"It's a solar system thirty-nine light years away. It has seven planets that revolve around an ultracool dwarf star. Some of those seven planets are similar to Earth in size and atmosphere. Some of them might very well host intelligent life."

Right. And the moon is made out of cheese.

"Why is it considered an 'ultracool dwarf star'?"

"Because it's considerably smaller and cooler than our sun. If our sun were the size of a basketball, the TRAPPIST-1 star would be the size of a golf ball. It's so small that a year on its closest planet is only a day and a half in Earth time."

"That's small."

"Yes sir. Nevertheless, it still generates more than enough heat to support life on nearby planets, particularly since those planets are relatively close to that star."

"How do astronomers know such a system even exists if it's so far away?"

"By way of powerful space telescopes and ground observatories. They think three of planets around the TRAPPIST-1 star have rocky terrain and oceans, just like Earth."

"So, you're saying our alien ship probably came from the TRAPPIST-1 solar system?"

"Probably might be a bit strong, considering that ship could have come from any number of solar systems. But TRAPPIST-1 is closer than most and one of its planets is particularly Earth-like."

"What's it called? Earth junior?"

"It's called TRAPPIST-1e. It receives almost an identical amount of light to that of Earth."

"Let's say our shuttle out in the Everglades came from TRAPPIST-1e. How long did it take to get here?"

"If it was traveling seventeen-thousand five-hundred miles per hour, or the speed of our space shuttles, it would take almost one-point-four million years. But the aliens have no doubt developed spacecraft that fly significantly faster than the speed of light."

And I no doubt will be the starting quarterback for the Miami Dolphins this season.

"How fast does light go?"

"More than one-hundred eighty-six thousand miles per second. If a spaceship could travel at the speed of light, it would circle the earth seven-point-five times in one second. In theory, nothing can travel faster than the speed of light. But that's human theory. Not alien theory."

"Yes. Of course. Alien theory."

"I detect skepticism, Geo."

"Look, Saul. It's just hard to believe. First, that scientists can even draw conclusions about a solar system that's so far away and second, that intelligent life exists on another planet."

"Well, my friend, you need to change your thinking. The universe has trillions of planets. You can bet your butt that an awful lot of those support life considerably more intelligent than that on Earth."

Geo thought about Sam and all her dumb news decisions. "Actually, that's beginning to sound plausible. Gotta run, Saul. I'm on deadline. Thanks for everything."

"If you get a chance, check out the moon tonight, Geo. It will be between the Procyon and Pollux star systems right after dark. Should be very pretty."

"Right."

Geo hung up and drummed his fingers on his desk. Finkelstein's insights were pretty far out there yet intriguing. He decided to take a risk and lead off the shuttle story with Finkelstein's assertion that it might be alien in origin. That was the most interesting angle, after all.

Settling at his computer, he started writing. He emphasized Finkelstein's premise, that many planets in the universe likely sustain intelligent life. He mixed in Bud Kowalski's information about the dust, indicating the shuttle had traveled deep into outer space.

To balance the story, he noted that federal authorities feared the shuttle potentially was a Russian-made, first-strike weapon. Although Sam had earlier cut that element, he felt it was too crucial to ignore. He knew she would scream at him for resurrecting it but didn't care.

Once done, he shipped the story to a production directory, where Sam could edit it. However, she wasn't in her office that afternoon.

"Your story in?"

Geo turned to see Sharon Qi, the online producer, a woman of Chinese descent, standing next to his desk.

"Yeah. Sam wanted it early."

"She's having lunch with Joe Cronk. She asked me to give it a read and post it online."

Geo knew that Sharon routinely edited stories at Sam's behest. He had no objection; Sharon respected his experience and he, hers. More importantly, she usually didn't screw around with his copy.

"No problem."

Sharon returned to her desk and went through his story line by line, mainly looking for typos or misspellings. She rarely questioned his facts or logic.

Yet twenty minutes later she approached Geo. "You really want to say that aliens might have flown that shuttle here?"

"Listen, I think that's bullshit too. But I got an expert saying it's a possibility. On top of that, the feds are completely bamboozled by its origin. So, yeah. That's the best angle."

"It *is* interesting though completely unrealistic. How do you suggest I play it online?"

Geo figured there wouldn't be any repercussions to the alien angle because the story was unlikely to be widely read. Nothing the *Banner* produced ever was. "Play it big. It's unquestionably the most interesting story of the day."

Sharon nodded. "I think you're right."

14

Sam and Joe Cronk, publisher of the *West Broward Banner*, enjoyed a leisurely lunch at an upscale restaurant at the Seminole Hard Rock Casino and Hotel in Hollywood.

He started with a dry vodka martini and feasted on a sixteen-ounce ribeye steak, served with mashed potatoes, green beans and a skewer of shrimp, while she had a glass of chardonnay and a Waldorf salad.

When they were finished, they walked to the parking garage and his $230,000 Bentley Bentayga SUV, parked next to her Jaguar. He turned on the engine and air-conditioning and, as usual, they slid into the backseat, where he had his way with her and she pretended to enjoy it.

She detested coupling in his vehicle and did so only because Cronk found it thrilling. Of some consolation, the windows were tinted dark, disallowing anyone from viewing in.

On this day, possibly because he had downed a second martini, he labored to finish, and she had to gut out his obnoxious grunting. She hated his raunchy breath and how his bristly mustache irritated her mouth and neck. It seemed to take forever before he finally released a satisfied moan. He pulled up his slacks while she pushed down her skirt.

She rested her head on his shoulder, both now sitting up. "I have to tell you what happened with George Leon this morning."

"What did the twerp do now?"

"He threatened to quit unless I give him more latitude, as he put it. Said he's ready to go flip burgers at McDonald's if necessary."

Cronk chuckled. "Let him go. I never liked the guy."

"I wasn't happy about his attitude. He just doesn't seem to get that we don't cover hard news. On the other hand, he's an excellent reporter and writer, and I don't want to lose him."

"So, you caved?"

"*Caved* is a little harsh, Joe. I met him halfway. I told him that if he gives my assignments priority, I'll allow him to write about that shuttle on his own time."

Cronk pushed her away. His gray eyes narrowed. "After I explicitly told you I wasn't happy about that first story, you're allowing him to do another?"

"Yes, I am. It's not like I'm going to give the story good play. But a shot of hard news wouldn't hurt, Joe."

"Like hell," Cronk spat. "We're not going to compete with television and the internet. Our formula of playing only local, happy stories is the reason we continue to make money."

"I know that, Joe. But Geo got some exclusive information that even the news stations don't have. I don't see any harm with letting him write it up."

Cronk pulled a Cohiba and a sterling silver lighter out of the inside pocket of his jacket, which was hanging on the backseat hook.

He lit up the expensive cigar and puffed clouds of smoke, something he commonly did after coupling with Sam and something that made her queasy. She started coughing, prompting him to crack a window.

"What were you saying?" he asked as he snapped the lighter shut.

"That Geo obtained exclusive information about that shuttle. I see no harm in letting him write hard news once in a while."

Cronk puffed on his cigar. "I do. Tell him to forget about that story. Better yet, fire him. I don't want him taking control of our product, the way we do things."

Sam leaned to the other side of the SUV, opened the window halfway and turned to Cronk. "Look, Joe. He may be a pain, but he is far and away my best reporter. We need to hang onto him."

"I don't care how good he is; if he's an attitude problem, you need to get tough and let him go. Otherwise, giving in to him will come back to bite you in the ass."

To emphasize the point, he reached a hand under her skirt and cupped the right side of her buttocks, the cigar protruding from his mouth, a contented look in his eyes.

Sam was suddenly reminded how wrong and disgusting it was to service the retired millionaire, who was old enough to be her grandfather.

She tried to convince herself that she was lucky to be desired by a man so rich and successful. She told herself that without Cronk, she wouldn't have a good job. She tried to remember that she was in debt to him for advancing her career.

Nonetheless, she was tired of submitting to him. She grabbed his forearm and wrenched his hand free of her rear end. "What excuse did you give your wife for being with me today, Joe? You tell her you're at an Elks meeting again?"

"I thought we were talking about Geo, not my wife," Cronk snapped. "What's your problem, honeybun?"

She hated it when he called her "honeybun." It was demeaning and condescending. "No problem, Joe. I just wish you'd shave that hair in your ears."

Cronk finally realized what was going on; she was angry and, for all he knew, on the verge of ending this arrangement. Puffing more smoke, knocking the ashes out his side window, he wasn't about to risk losing his young mistress.

"Okay. I hear you," he relented. "But don't let this get out of hand. Let him do a shuttle story. A *little* shuttle story. Then make sure he gets back on your schedule."

Despite the cigar, she put her head back on the old man's shoulder, knowing she had won a victory. "Thanks. Why don't you like him, Joe?"

"I know his type. He's a good-looking guy who thinks he's entitled. He expects everything to be hand delivered without having to work for it, be it women or a paycheck."

"Sounds like you're jealous, Joe."

Cronk maneuvered the cigar to the side of his mouth. "When he's made his first five million, like I have, then I'll be jealous."

Sam looked out the backseat window and saw people coming from and going to cars in the Hard Rock's massive garage, many of them chronic gamblers. Cronk was wrong to think Geo was a slacker, she thought. He was a hard worker who expected nothing special in return.

But Cronk was right about one thing: Geo was indeed a handsome man. In her estimation, *very* handsome. For that reason, she wanted to keep Geo around for more than his journalistic skills. The truth was she secretly lusted for him and couldn't wait to get back to the office to see him – even though their relationship was a bit rocky at times.

First, however, she wanted compensation for satisfying the publisher and dealing with his pungent smoke. "Joe, I want to buy some shoes and a dress I saw at Dillard's. I also want to get my mother some perfume she likes."

"How much you need?"

"Two thousand should do. Make it three."

Cronk could have questioned why shoes, a dress and perfume cost so much. He could have chastised her for taking advantage of him. Instead, he reached into his jacket and pulled out his checkbook and a Montefiore pen. Nonchalantly puffing on the cigar, he began filling out a check.

"No problem, honeybun."

15

The skies were gloomy gray on Friday morning, the fourteenth day of February. Thanks to a strong cold front, temperatures plunged into the low fifties, chilly by South Florida standards.

But it was warm and friendly inside the Hitching Post, a popular restaurant in downtown Davie, where Geo took Gina for Valentine's Day breakfast. The place was full of couples, chatting, laughing, holding hands.

A waitress dropped by their table to pour coffee and take their orders. Both asked for scrambled eggs, bacon and toast.

"This already has been the best Valentine's Day in my entire life," Geo said, checking his watch. "And it's not even eight o'clock."

Wearing one of Geo's sweatshirts, she reached a hand across the table to hold his. "You sure had a lot of energy last night. This morning, too."

He smiled, recalling how sweet both sessions were. His angst over a fraudulent marriage was steadily being replaced by a legitimate desire to spend time with her.

"Like I said, this has been one hell of a nice Valentine's Day."

She sipped her coffee, holding the cup with her free hand. "What is Valentine's Day, again?"

"They don't have Valentine's Day in Mexico?"

"Not in my part of Mexico. Is it like *Cinco de Mayo*? An excuse to get drunk?"

"Not exactly. It's a day when a man is required to do something nice for his wife or girlfriend."

"You consider me a girlfriend?"

He nodded. "We haven't known each other long, but, yes, you're pretty much my significant other now."

She gave him a quizzical look, and he thought she was going ask him what that meant. Instead, she leaned over the table, grabbed a fistful of his shirt collar, pulled him close and gently kissed his lips. An elderly couple watching from a nearby table smiled.

"Significant other. I like that."

The waitress soon delivered their breakfasts. They released hands and hungrily attacked their plates. When they were finished, Gina put down her knife and fork and asked, "Do you believe in God?"

Geo dabbed his mouth with a napkin. "What was that?"

"You heard me. Do you believe in God?"

"Why're you asking?"

"Because we're about to get married. We might spend our lives together. I want to know your beliefs. Do you believe in God or not?"

Shit, he thought. Just when he was warming up to the idea of tying the knot, she turns out to be a religion fanatic? Maybe a Jesus freak? Yet he noted she wasn't wearing a crucifix.

"The truth is, Gina, I'm not very religious. I'd have to classify myself as an agnostic. I'm not sure if there's a god or not."

"You're not sure," she repeated slowly.

"Correct."

"So, you never pray to God." It wasn't a question, more like an accusation.

"Sometimes, when I'm in a tight spot, I'll say, please God, don't let this happen or please make that happen. But it's just an expression. I don't actually believe there's some supreme being who will intervene on my behalf."

"You don't think God can save you. Or that God loves you. Right?"

"Right. If I need to be saved, I'll try to do it myself."

"Good luck with that."

Where's she going with all this?

"How about you? Do you believe in God?"

"I don't *need* to believe in God. I *know* God exists."

"I see," Geo said, his fear resurrected that something was off about her beyond her propensity to live dangerously. "You're sure, then, that he'll save you?"

"No, I'm not sure. But I hope he will with your help."

"How in the world could I help you?"

She put down her coffee cup. "You can accompany me to God's house."

Geo smirked until he saw that she was dead serious. Okay, she's illegal, she's a God freak and she's whacko, he thought.

And I had to go and fall head over heels for her.

Just then a huge man exploded through the front door of the restaurant, accompanied by two buddies, a mishmash of tattoos, scuzzy beards and bandanas.

Geo immediately recognized him as the biker who had been hunting for Gina after she stole his cycle. Gina saw him too but seemed unconcerned. She held up her cup to the waitress for a refill.

The big goon looked around the room, clenching his fists. Spotting Geo and Gina, he stomped over to their table, his two accomplices in tow, their leather boots thudding against the wood-plank floor. The waitress, who was about to pour coffee into Gina's cup, instinctively backed away.

"Finally got you where I want you, bitch," the biker growled at Gina. "ICE ain't gonna save your ass this time."

With the swipe of an arm, he knocked cups, plates, silverware, salt and pepper shakers and a bottle of Tabasco off the tabletop. It all crashed to the floor, startling other patrons.

Geo jumped to his feet, alarmed. The restaurant manager, a middle-aged woman who had been near the cash register up front, pulled out her cellphone to call the police.

Gina seemed oblivious. "Good morning, Buzz."

"Not going to be such a good morning for you, you fucking bitch."

Grabbing her upper arm, he started to tug her out of her seat. But Geo shoved the biker as hard as he could, much as he had done with the lecherous thug at Shorty's.

"Leave her alone, you stinking ape," Geo spat.

Buzz released Gina and glared at Geo. "That wasn't smart, you little fuck."

"You wouldn't know smart if it walked up and kicked you in the teeth. Why don't you and your cronies take a hike?"

Buzz grinned, revealing tobacco-stained, crooked teeth. "Now I remember you. You're the one what said, if I want to get to her, I'd have to kill you first. That right, you little fuck?"

Out of the corner off his eye Geo saw Gina waving at him to sit down.

"That's right, lamebrain."

"Lamebrain, huh? You little fuck. I'm gonna hurt you bad." The biker turned to Gina. "Then I'll take care of you."

He was just about to smack Geo when Gina got up.

"Wait, Buzz. You're right. I need to pay for what I did. Let's go to the men's room. I'll make things right."

"What?" Geo bellowed. "Like hell you will!"

She calmly looked at him. "Just stay here, Geo. This is something I have to do."

"Damn straight," the big biker declared.

Gina looked at his buddies. "Bring them with you."

Buzz shrugged. "Fine by me. I like foursomes."

"Righteous!" one of the other bikers blurted.

Ignoring Geo's bewilderment, Gina walked towards the men's room, Buzz and his two cohorts close behind. Numerous diners watched them, most amused. The restaurant manager remained by the cash register, anxiously looking at her watch, waiting for the cops to arrive.

Before going into the restroom, Gina looked back at Geo, who was still standing next to their table, and winked at him. Then Buzz shoved her through the door, and the three men followed.

Geo sat down and placed his head in his hands. He was debating whether to try to somehow stop whatever was happening in there or just leave the restaurant and outright forget about Gina.

Do I really want a woman who's voluntarily satisfying three thugs?

Leaning toward the latter option, he was about to get up when Gina came out of the restroom alone – less than two minutes after she went in. She walked back to the table, seemingly untouched.

"What happened in there?"

She sat down. "I sure as hell didn't fuck them, if that's what you're asking."

"Are they still in there?"

"Yep."

"Mind if I go take a look?"

"Be my guest."

Geo all but ran to the men's room, passing tables full of equally curious patrons. Inside, he didn't see them at first.

Then he noticed a stall door was partially open. Peeking inside, he saw Buzz seated on the commode, head back, eyes closed, arms sprawled to the sides, his pants around his ankles. The other two were on their knees, their heads in Buzz's lap. All appeared to be unconscious.

Geo put two fingers against Buzz's neck to make sure he was still alive. Relieved that he was, Geo guessed that Gina had tasered the three, the only reasonable explanation.

He walked back to the table while people whispered, trying to figure out what was going on. Gina was sipping coffee as the waitress finally gave her a refill. He didn't sit down.

"How did you do that to those men so fast?"

"I'm stronger than I look. I know how to fight."

"Did you taser them? Then positioned them like that?"

"What's the difference? They got what they deserve."

Suspecting the cops would arrive any second, Geo threw enough cash on the table to cover the two breakfasts, the broken plates and glasses and a tip. "Let's get out of here."

"If you say so."

They pulled out of the parking lot just as five police cruisers screeched to a halt in front of the restaurant, red and blue lights flashing. Gina giggled as the officers rushed inside.

As they drove to his townhome, Geo continued pondering how she had so quickly dealt with those animals. He hadn't forgotten that before that incident, she wanted him to accompany her to "God's house." What in the hell did she mean by that?

"Don't worry, Geo. I'll tell you all about God's house when the time is right."

He whipped his head to her, amazed that she always seemed to know what he was thinking. "How did you know I was –"

She took his hand. "You're easy to read."

When they arrived at his townhouse, he followed her inside and closed the door. She put her arms around his neck and whispered in his ear.

"You tried to save me again. You think I'd just forget about that?"

Though he was dying to take her right there, to revel in the complete satisfaction that he now knew she could deliver, he took her by the shoulders. "I have to go to work."

She slowly nodded, wisps of hair covering one of her eyes. "I'll be waiting for you when you get home."

16

As he parked in front of the *West Broward Banner*, Geo knew it was going to be a long day of dealing with Sam's bullshit and trying to decompress after the scene at the Hitching Post.

Although Gina had saved him from the big biker's wrath, it rankled him that she had so nonchalantly subdued the thugs in a mysterious way. What happened in that restroom, anyway? He guessed he would never know.

Sighing, he went inside. Before he reached his desk, Sam came out of her office.

"Geo, get in here. Pronto."

Christ, what did I do now?

He stepped into her office. "Good morning, Sam."

"Close the door," she snapped. "Sit down."

He sat down in front of her desk while she fell into her executive chair.

"What's the problem?"

"Your story has put us in the middle of a shitstorm. That's the problem."

"How so?"

"Because," she seethed, "the story stated that extraterrestrials might have flown that shuttle here from another planet. How could you be so reckless?"

He furrowed his brow. "That was hardly reckless, Sam. I didn't outright say aliens flew it here; I said they *might have*. That came from a credible source, Saul Finkelstein. When it comes to astronomy, he's an expert."

"Oh, really? That's not what the *New York Times* says online today. They wrote a *whole fricking story* in response to yours. They say NASA fired Finkelstein after he attempted to stow away in a space shuttle. They say his coworkers considered him a nutcase. Did you even check him out before you used him as a source?"

"I didn't need to check him out, Sam. I was aware of the stowaway thing. I also know he's a bit eccentric. But this guy used to be a rocket scientist. If my story was so sensitive, why didn't you read it before it was published?"

"Don't turn this on me! I let Sharon edit it, as I often do. I just didn't expect to be broadsided by such foolishness. Now we're a laughingstock. By the way, Sharon is in a boatload of trouble too for letting the story go."

"Don't pin any of this on Sharon. It was my decision to run it. Her job was to edit and post it."

"Her job was to exercise news judgment."

Geo almost laughed. *What the hell do you know about news judgment?*

"What media are ridiculing us?"

"All of them! The *New York Times*, the *Associated Press,* the *Washington Post* and a bunch of others. All posted stories this morning, ripping us. Even the networks are calling us idiots!"

"The networks?"

Using a remote, Sam turned up the volume on her television, which was tuned to *Good Morning America*. "Perfect timing," she muttered.

"As we've been reporting all morning, it seems our planet is being invaded," Michael Strahan said, trying to keep a straight face. "According to a Florida newspaper, aliens from outer space landed that shuttle in the Everglades, evidenced by strange dust all over it."

Co-host Robin Roberts pretended to be alarmed. "Oh my. I hope this doesn't end like that *Twilight Zone* episode where the aliens take Earthlings back to their planet and cook them for dinner."

"I'm sure they'd spit me out," Strahan said. "I don't taste that good."

"With some salt and ketchup, I suspect you'd make a very savory, filling meal," Roberts countered.

"Is that a shot at my weight?"

Roberts laughed. "Lay off the Snickers, Michael."

Meanwhile, the movie *Aliens* was playing in the background, showing a drooling, slimy creature.

Sam switched to the *Today* show, where co-hosts Savannah Guthrie and Hoda Kotb were seated with weatherman Al Roker, who was wearing a Darth Vader mask.

"Darth, is it true that you flew that shuttle to Florida?" Guthrie asked him.

Roker responded in a fake, low voice. "Yes, it is. Delta and United were booked, so I had to shuttle in."

Chuckling, Guthrie inquired, "What are you planning to do during your stay here?"

"Take over your planet, of course," Roker replied. "That is, if the weather holds up. We have a cold front moving in from the Midwest, and it's expected to bring rain to ..."

Sam switched channels again, this time to the *CBS Morning Show*. Gayle King, one of the anchors, was interviewing a Harvard astronomy professor.

"It's fun to speculate that the spacecraft might be alien in origin," he was saying. "But if that were the case, that ship would have come from another solar system, far beyond ours. It would be difficult for any species, no matter how advanced, to survive such a long trip."

"Which is to say, what?" King asked.

The professor looked directly at the camera. "The story in that Florida newspaper is a bunch of horse manure."

Sam switched off her set and pointed an angry finger at Geo. "Thanks to you, we're a punchline!"

"Sam, I wrote the piece based on the best information I had at the time and – "

"Our own readers are disgusted. Look at this!" She turned her computer screen, allowing him to see her inbox was teaming with emails. "Most of these say it was irresponsible for us to run that story. Many readers are threatening to cancel their subscriptions."

"Sam, I –"

"Then there's Twitter, Facebook and Instagram. You should see the nasty shit people are saying, calling us dimwits and idiots."

"Christ. Since when have extraterrestrials become so controversial?"

"It's not a matter of controversy. It's a matter of credibility, Geo. Because of you, we lost ours."

He understood that his story had stirred up the pot. Yet, if anything, he thought Sam should thank him. He had just put the small, irrelevant *West Broward Banner* on the national map. But she wouldn't see it that way.

"So, what now? You going to fire me?"

"I should. But I'm going to give you a second chance – under two conditions."

"Which are?"

"First, you write a correction saying your story erroneously made it sound like it was a fact that aliens flew the shuttle here, when we should have said it was one nutty astronomer's theory. Second, and I mean *today*, you do a feature on the Davie Knitting Circle, a group of women who meet every week to knit sweaters, quilts and the like."

Geo glared at her. "You have got to be kidding."

"Trust me, I'm not."

"Sam, I can't write a retraction on a story that was properly reported and, as unlikely as it may seem, could be true. As for the knitting circle, blank space would be more interesting."

She leaned forward. "Don't you dare argue with me, George Leon," she hissed. "Do what I tell you or else hit the road. Your choice."

Geo glanced at the door. He could have quit in a fit of indignation. But he wasn't a quitter and, more importantly, wanted to justify what he had written about the shuttle.

"I'll have the retraction to you in fifteen minutes," he grumbled, getting up.

"One more thing. Don't write anything that dumb ever again."

Oh. And the Davie Knitting Circle isn't dumb?

"Right."

After Geo departed, Sam leaned back, closed her eyes and pinched the bridge of her nose. Although agitated, no way was she about to fire Geo, not when she frequently fantasized about being intimate with him, even as she pretended to be busy in her office.

As a manager, she knew it would be risky to come on to him. That could be construed as sexual harassment. All she could do was hope that Geo would make the first move, even though he hadn't shown any interest in her thus far.

In the meantime, she would continue to sneak glances at him and envision their bodies meshed together in the utility room, in the bathroom, on top of her desk. Such fantasies helped her endure the icky sessions with Joe Cronk.

She was thinking about the old publisher, wishing he would just go away, when her phone rang. "Speak of the devil," she mumbled and picked up. "Hey, Joe. What's up?"

"I told you that moronic reporter would bite us in the ass, and Christ, did he ever."

"I know. I've been monitoring all the reaction."

"How in the hell did it get published, anyway?"

"Sharon put it into production without consulting me."

She didn't mention that she had given Sharon authorization to decide what stories would run because Sam was uncomfortable making those decisions herself.

"Then fire her and Geo at the same time. When I told you before to fire him, you stuck up for him. Look where *that* got us. Now, be a good girl and kick both their asses out the door. Then do some damage control. Send out a press release letting all the other media know we took corrective action."

Sam was quiet. Then she resolutely uttered, "No."

"What do you mean, 'no'? I'm giving you an order, Samantha."

She hated being called by her proper name. It was almost as bad as honeybun. "And I'm telling you, no. I'll take the blame for not checking what stories Sharon had slotted and for not reading Geo's story before she did."

"Nice of you to fall on your sword. But you delegated authority to Sharon and she blew it, placing the story at the top of the homepage. I want both of those people gone, Samantha. Today."

"Again, no, Joe. I need both of them. If you really want to fire them, then you have to fire me too."

It was the first time she had stood up to him so firmly, and it felt good, even though she was putting her neck on the line.

Cronk briefly considered being a ruthless bastard and telling Sam to pack her desk. But that would mean losing the best thing in his life.

"Okay. If that's how you want to play it. They can stay for now. But you should know: I'm not happy."

"Duly noted."

Neither one said anything for a few seconds, an awkward silence. Then Sam heard him snap open his lighter, fire up a cigar and take a puff.

"See you tomorrow? At the hotel?"

She breathed in deeply. "I'll be there."

17

The Everglades were quiet on Saturday morning. No airboats were roaring through the marshes. Frog-giggers were done for the night. Fishermen had yet to arrive. Even the wading birds and alligators seemed to be asleep.

Bud Kowalski sat in his Ford Expedition parked at Dade-Collier Training and Transition Airport, not far from the mysterious space shuttle. He was sipping coffee purchased from a 7-Eleven, admiring the raw beauty of the Everglades and stewing.

His wife had been out all night again, slipping into bed beside him before dawn, reeking of alcohol and another man's cologne. Though she was blatantly cheating on him, he wasn't about to confront her again.

The last time he did so, he almost lost her. That would have shattered his world. He still loved her too much to break up their marriage of twenty-seven years. Further, they had two college-age daughters, and he didn't want to rip apart the family.

As gentle sun filled the sky, he again resolved to endure her drinking and cavorting. He again pretended everything was hunky-dory, even though, in reality, he was suffering. At least his work helped him keep his mind off of her.

A seasoned accident investigator, Kowalski had been instrumental in unraveling several high-profile disasters, most of them jetliner crashes in Asia and Africa. The spacecraft sitting fifty yards away was proving to be the biggest challenge of his career. Nothing about it made sense.

There was no evidence the spaceship even made a normal landing. Investigators found no fresh rubber marks on the runway from its wheels making high-speed contact, and no one witnessed its approach.

Kowalski had chuckled when he read Geo's story about the shuttle possibly being flown in by aliens. Though he didn't believe that, he sensed something extraordinary had happened here. He couldn't rule out that someone, other than the Russians, was involved.

One possibility: An emerging nation, heretofore ignored for its technical achievements, had launched the thing. Since the 1960s, nine nations had developed orbital spaceflight capability, among them Japan, Israel, India, Iran and North Korea.

Another possibility: Two or more nations had banded together to develop the shuttle. For instance, the European Space Agency was formed to advance space travel and included twenty-two members, including France, Germany and the United Kingdom, all known for their aviation expertise.

While none of this seemed likely, Kowalski knew nothing could be ruled out. Finishing his coffee, he got out of his SUV and checked in with the military guards posted near the shuttle. Then he yet again searched for clues that might reveal its origins.

He was about to inspect the spacecraft's underbelly when his phone rang. It was his wife.

"Hey, Emmy. What's up?"

He heard her sniff. "I'm moving out, Bud." He shut his eyes tight as he held the phone to his ear. "Bud? Did you hear me?"

"Yeah. I heard."

More sniffling. "I can't do this anymore, Bud. It's not fair to you. It's just not right."

After twenty-seven years, it sure the hell isn't.

"If that's what you want, Emmy, I won't stop you."

"It's not what I want, Bud. But it's what I need to do. I'll be gone before you get home tonight. I want you to know I honestly love you. It just isn't working anymore between us."

He opened his eyes and ended the call. Feeling blindsided, he wanted to get in his vehicle and drive for hours, long enough to dampen the pain she had just inflicted.

But he had a job to do, and Kowalski never shucked responsibility, no matter how tough the going got. Running his fingers over the exterior surface, he commenced a methodical inspection of the shuttle, using it as therapy to keep his mind off of her, to prevent breaking down.

After conducting this painstaking examination for more than an hour, he found something he hadn't seen before. Though it blended in almost perfectly with the surface, he noticed a touchpad below the cockpit window. It was similar to the pad on a laptop, allowing one to move the cursor with the touch of a finger.

He gently tapped the pad. It lit up. Feeling a bolt of excitement, he touched it again, hoping it might open a hatch. Nothing happened.

He continued touching the pad in different areas until he heard a sound inside the craft. It was a melodious yet eerie chime, produced by some kind of highly synthesized instrument.

He tried to duplicate the touching pattern. Again, no hatches opened. He could only surmise that a specific series of touches in the right places would unlock the shuttle. In any case, combined with the ship's advanced design, the pad indicated this was an elaborate piece of machinery.

He started to amble back to his SUV and froze in his tracks, contemplating an outlandish possibility. As incredible as it seemed, perhaps this spacecraft was not of this Earth, as Geo's story postulated.

Wheels started churning in his head. He knew he would be putting his professional credibility on the line to even suggest this ship was alien in origin. Nonetheless, he decided to see what top astronomers and physicists thought.

He walked briskly to his Ford Expedition, jumped in and, even before starting the engine, commenced making calls to line up experts. Soon he was on the Tamiami Trail, charging back toward Miami, growing more enthused about his crazy plan.

He thought about Geo. His journalist friend would no doubt be excited to hear that a bunch of eggheads were about to weigh in on whether his alien story might be true. Despite his broken heart, Kowalski smiled.

Coincidentally, Geo was already excited as Gina had climbed on top of him, pulling him out of sleep. "Again?" he asked groggily.

"I'm not done showing how grateful I am," she whispered in his ear.

"Sounds good to me."

The previous evening their lovemaking had been risqué and playful as they had frolicked in virtually every room in his townhouse. They had even hit his laundry room, where she had positioned herself on the dryer and turned it on.

They did take time to have dinner. He had whipped up scrambled eggs and poured two glasses of merlot. They ate this romantic meal straight out of the frying pan while sitting at his kitchen table naked. Now here they were, moaning and groaning again on a cozy Saturday morning.

When they were done, enjoying the afterglow, he was suddenly apprehensive. It occurred to him that he was hopelessly in love with a woman who might end up ditching him, much as Katrina had.

"I'm not her," Gina said softly, lying next to him.

Geo gazed at her, befuddled. "How in the world ...why would you say that?"

"I could tell you were thinking about her. Just the look on your face."

"You're pretty damn perceptive."

"Get something straight, pretty boy. I'm not Katrina. I'm not stupid enough to lose you like she did. You're my man, I'm your woman, and that's how it's going to be. We're together now."

He was elated to hear that declaration. He was also amazed that she had seemingly read his thoughts again. How did she even know about Katrina? He had never even talked about her. He figured that Gina must have found an album containing photos of him and Katrina.

"That's good to know."

She kissed his neck. "Last night, after you fell asleep, I was watching an interesting TV show."

He remembered the television had been on while they were entwined. She must have turned it off after he passed out. "What was it about?"

"A man named Joel Osteen was talking to an auditorium full of people."

"He's a televangelist. I've heard his sermons here and there. What did Pastor Osteen have to say this time?"

"That God created humans before they even arrived on Earth and called them a masterpiece."

"Why is that interesting?"

"Because Osteen got it right. God did create humans long before they surfaced on Earth."

Geo again feared she was a religious fanatic, having earlier questioned him about his beliefs. At that point, however, he didn't care if she was a raving zealot. He wanted to be with her, maybe for the rest of his life.

"You know this how?"

"I just do. I'm hungry. Let's get breakfast."

"Okay."

Leery of returning to the Hitching Post, they drove to a Denny's near Sawgrass Mills Outlet Mall and sat side by side in a booth. After expending so much energy in their lovemaking, both hungrily consumed a stack of pancakes smothered in butter and syrup, accompanied by orange juice and coffee.

"These are heaven," she moaned as she worked on the flapjacks.

He nodded but was looking around the restaurant. "They sure are," he quietly agreed.

"Don't worry. Those bikers aren't here."

He gawked at her. "Gina, how in the hell –?"

"It's a gift, sensing what's in people's minds. I can deal with those lunkhead bikers. The immigration guys are another story. Even if we get married, they might deport me back to Mexico."

"I won't let that happen. We're getting married. For real. Like you said, we're together now."

She put down her fork, put both arms around him his neck and hugged him. Just then, Bud Kowalski called.

"Got a development you should know about, Geo."

"What's that?"

"I'm asking topflight astronomy, astrophysics and earth-science experts to examine the space shuttle."

"For what purpose?

"To determine where it came from. I'm convinced the Russians had nothing to do with it, and all the private companies that design spacecraft categorically deny having anything to do with it."

"So, what are you saying?"

Kowalski sucked in a breath. "I'm saying we need to explore the possibility that this thing is extraterrestrial in nature, like your story said."

Geo sat up straight. All the other media had guffawed at his story. Now here was Kowalski, a credible government official, saying it might have been correct.

"So, these experts might be able to determine if the craft is from outer space?"

"That's about what it comes down to. That's hard to believe, but at this point we're at the end of our rope and need to check all possibilities."

Geo looked upward as though thanking God. Even if these specialists concluded aliens weren't involved, just the fact that they were checking was vindication enough – after his story was roundly trashed.

"When will the experts start their inspection?"

"Soon as they all get here. Once they're done looking the thing over, we'll compile their findings in a report."

"Let me guess. That will take weeks."

"I hope days."

"Can I watch them work?"

"Sorry, no. All media will be prohibited from entering the airport property while they're here. We don't want the experts to be distracted. But I'll call you when they're done."

"Okay. Thanks for the heads-up, Bud."

He put down his phone and pulled Gina close.

"What was that about?" she asked.

He hadn't told her about how his shuttle story had pissed off his editor, almost resulted in his dismissal and drew mockery from the other media.

"Just a source basically confirming the shuttle out in the Everglades wasn't built by the Russians."

Gina looked at him curiously. "What shuttle?"

18

When they returned home, Geo didn't park the Jeep. "I'm heading to the office. I have to do some work," he said, remaining behind the wheel.

"But it's Saturday."

"I know. But I've been working on weekends my whole newspaper career. It's the only way to get ahead of the game."

"In that case, go."

"I'll be home this afternoon. See you later, Gina."

Still in the passenger seat, she kissed him. "You'll see all of me later," she cooed and held out her hand. "Give me your credit card. I need to buy some stuff."

He pulled a Visa out of his wallet and handed it to her. "Have fun."

She slid out of the Jeep. "Always do."

After Geo drove away, Gina mounted her stolen motorcycle and rode to a Walmart on Flamingo Road. She wanted to buy some lingerie to make it an exciting afternoon.

Inside the expansive department store, she went to the women's section. She was inspecting a red thong when she felt a presence. She turned to see a man standing a few feet behind her, staring at her rear end.

Stocky with a beer gut, he wore a black T-shirt emblazoned with the bright yellow words, "Dogs Die in Hot Cars," which actually was the name of his favorite rock group. His khaki shorts exposed flabby legs. His face was a raunchy mix of coarse black stubble and pockmarks. He looked as though he hadn't bathed in days.

Ignoring him, she moved to another clothing rack and began checking out bras. Then she noticed the creep was closer yet, videoing her with his smart phone.

Sighing, she turned to him. "What's your name?"

He hesitated a moment, wondering if he had targeted a cop. "Gus," he finally answered.

She plucked a pair of oversized panties, some pantyhose and a women's belt off nearby racks. "Listen, Gus, why don't you come with me to the dressing room. You can watch me try this stuff on."

A lecherous smile spread across his ugly mug. His voice was raspy. "Oh, yeah."

He followed her to the dressing room area, where a woman attendant put up a hand to stop them. She regarded the weirdo behind Gina.

"Everything okay here, ma'am?"

"Yes. He's with me. He's my boyfriend."

The attendant gave her a look. *You're kidding.* Shrugging, she pointed toward a room at the end of a short hall. Once Gina and Gus were inside, he locked the full-frame door, which had no viewing space above or below it.

He turned to her, his lewd grin revealing spaced out, decayed teeth. His rancid breath and sweaty body odor quickly dominated the small space. He took a step toward her. She took a step back and found herself in a corner.

He wagged his tongue at her. "I want to taste you. Down there."

He came closer and was about to start groping her. Suddenly his hands, arms and his legs went dead. He was unable to move a muscle, paralyzed.

She snickered. "You should be more careful who you mess with, Gus."

A minute later Gina walked out of the dressing room alone and returned to shopping. She selected a matching red bra and panties, a garter belt and fishnet stockings. She also grabbed some blue-jean cutoffs. Using Geo's credit card, she paid for the items at the front of the store.

It was only after she departed that the woman attendant found the chubby oaf still in the dressing room, lying on his back, naked and whimpering. He had been hogtied with a belt, all four of his limbs clamped together. Oversized panties had been stuffed down his throat and secured with pantyhose wrapped around the bottom of face.

His smartphone was on the floor by his head, all videos of Gina deleted. But videos of other unsuspecting women remained in the phone's memory. The attendant called the store manager, who called the police, who took the unkempt man to their station and questioned him extensively.

Despite claiming he was attacked by a "flaming bitch," the man was charged with voyeurism based on images found on his phone. Gina, meanwhile, gleefully gunned the motorcycle back to Geo's townhouse.

The lingerie worked exactly as she intended. The bra, panties and garter belt ended up on the floor of Geo's bedroom that afternoon. Exhausted and content after their lovemaking, they lazed in bed, their backs against the pillows.

"Gina, while I was at the office this afternoon, I made arrangements for us to marry at the courthouse on Monday. Our deadline, per Bob Klug, the guy from ICE, is Tuesday."

"Okay. That is, if you really want to marry me. If you don't, it's not too late for me to ride out. Don't do anything you don't want."

He rolled off the bed and pulled her to her feet. As both stood naked, facing each other, white carpeting under their toes, he dropped onto one knee.

"You're not riding anywhere, and this is something I want to do. In the short time I've known you, I've fallen madly in love with you, Gina Martinez. Will you marry me?"

She remained stone-faced for a couple of seconds. Then she smiled. "Yes."

Pulling him up, she pushed him back onto the bed. Though she never said as much, he figured it was her way of saying she loved him as well.

Mellow afternoon sunlight was seeping through his window blinds when his text-message bell rang, awakening both of them. He grabbed his phone off the bedstand and read the note.

"Who's that?" Gina asked, without lifting her head from her pillow.

"Saul Finkelstein, the guy who told me extraterrestrials might have flown that shuttle out in Everglades here."

"He honestly thinks that?"

"Yep. He sure does."

"What's he want?"

"He's holding a lecture at Markham Park tonight. He thinks I should be there."

"Why?"

"He claims aliens have arrived on Earth."

Gina propped herself up on an elbow. "Let's go. I'd love to hear what the guy has to say."

"I would too."

19

Samantha Pauline Vandermark and Joe Cronk were going at it in the Courtyard Marriott next-door to Westfield Mall. At least he was. She was staring blankly at the ceiling, his face nestled into her neck.

As the hotel was near the busy intersection of Pine Island and Peters roads, she could hear the whoosh of afternoon rush-hour traffic outside their second-floor room. The only other sound she heard was Cronk's pathetic grunting, which reminded her of a bleating sheep.

For him, it was a naughty pleasure to cheat on his wife with a pretty girl almost forty years his junior. This, even though his wife was a beautiful woman in her own right and twenty years younger.

For Sam, it was drudgery, not to mention demeaning. She felt cheap and used and was growing tired of this game. Yet she had only herself to blame. She had willingly agreed to this arrangement in the name of fostering her career.

They had met two years earlier at a journalism job fair, held at a Miami hotel. The fair allowed newspaper editors to assess the latest crop of journalism school graduates. She had been fresh out of Florida State University with a master's degree and big dreams.

Originally from Fort Lauderdale, raised in a middle-class home, Samantha Pauline Vandermark wanted to be an editor ever since she spent two years working for her high school newspaper, reporting stories, writing editorials and selling ads.

Juiced by the lively environment of the newsroom, she majored in journalism at the University of Florida and became a copy editor for *The Alligator*, the school newspaper.

Despite her enthusiasm, Sam's professors considered her middle of the pack. One faculty member bluntly told her: "Sam, you might want to consider becoming a model or an actress. Journalism isn't your strong suit."

Undeterred, she worked even harder at writing and editing and won a full scholarship to attend Florida State University. There, she earned respectable grades and continued her quest to be an editor.

However, because newspapers across the nation were struggling, she knew landing an entry level position, let alone an editor job, would be difficult.

Thus, when she attended the job fair in Miami, her goal was simply to get an interview with a newspaper – *any* newspaper. She was elated when Joe Cronk, publisher of the *West Broward Banner*, chatted at length with her and urged her to submit an application.

Cronk, meanwhile, had been filling in as editor of the *Banner* after firing the previous editor. With no journalism skills whatsoever, it was a job for which he was ill-equipped. He was anxious to hire a real editor.

Prior to buying the small newspaper, Cronk had been a top executive for a big automaker. Although paid a seven-figure salary, he hated the long hours and stress. So, he retired to South Florida and bought the *Banner* to keep busy.

During their informal chat at the job fair, Sam had been impressed by Cronk's wit and charm. She also found him rather handsome for a man in his mid-sixties. At the conclusion of their meeting, she invited him to have a drink at the hotel bar.

She wasn't thinking in terms of a sexual rendezvous. Rather, being ambitious and outgoing, she wanted to be on friendly terms with a man who might be her future employer. She was hoping he would offer her a job as a reporter or a copy editor.

Yet, after a couple of drinks, Cronk suggested she could start right away as *editor* if she accompanied him to his hotel room. He didn't have to spell it out. For such a professionally rewarding offer, particularly for such a young, inexperienced woman, the unspoken implication was that she would have to accommodate him.

Thinking of the prestige of being an editor, she didn't hesitate to follow him up to his room and undress. She knew it was wrong. But she told herself: *Get over it. This is the price for a break.*

True to his word, Cronk named her editor of the *Banner* and paid her a generous salary. Then he regularly asked her to meet him outside of the office, sometimes in his vehicle, usually at hotels. She knew this was classic sexual harassment. She also knew she was degrading herself. But she never said no.

Filled with more exuberance than morality, she wasn't about to pass up an opportunity to gain valuable experience as an editor and potentially advance to a major newspaper, her ultimate goal.

In return for this unscrupulous deal, she was given authority to run the *Banner* as she desired. She hired reporters, selected stories and determined how the front page would be designed.

Yet from day one she was in over her head. Thanks to her inexperience, she lacked the basic skills to be an effective editor, including news judgement, time management, scheduling, and financial planning. The only thing she did well was boss around her employees.

Despite her ineptness, the *Banner* still flourished mainly because Cronk insisted it cover highly local and happy news. He knew full well that hard news was becoming a thing of the past in print publications.

Because Cronk rarely showed his face in the newsroom, the staff had no idea the publisher was screwing the editor. Only Geo knew about their connection after secretly observing them on her office couch that Sunday night.

That was the only time Sam and Cronk dared to meet at the newspaper. She otherwise wasn't about to invite Cronk to her mother's apartment, where she lived, and he couldn't very well bring a young woman into the home he shared with his wife.

So, here they were at the Courtyard. Although clean and comfortable, Sam hated the place. For it was here that she most often had to endure Cronk's pathetic sexual routines, his cigar-stench breath, his scratchy mustache and that horrible grunting.

She managed to get through these sessions only by fantasizing that she was with another man, and invariably that was Geo. She would close her eyes and imagine he was making passionate love to her. Occasionally, she even felt pleasure.

Such was the case on this afternoon, when she inadvertently whispered, "Oh, Geo," into Cronk's ear.

He rolled off of her. "You're thinking about him? While you're with *me*?"

She inwardly cursed herself. "Don't make a big deal, Joe. He's been on my mind because of things at work. That's the only reason I said his name."

"Bullshit," Cronk spat, getting out of the bed. He went to the chair, where he had neatly assembled his clothing. "You hate being here with me, I can tell. You wish you were with him," he grumbled, pulling on his boxer shorts.

Sam sat up, exposing her bare breasts. "That's not true, Joe. I was thinking about his next assignment when my mind should have been on you. I've been under a lot of pressure lately, particularly after all this space shuttle business."

Cronk put on his trousers. "Again, bullshit. I wasn't born yesterday, Samantha. I know this isn't a pleasant situation for you, but I've tried to be good to you and honest with you. You, unfortunately, can't say the same."

She got out of bed naked and put a hand on his arm. "Joe, I know you've been good to me, and I'm grateful. I try to prove that each time we're together. I'm here because I want to be with you."

Which was another lie, and Cronk knew it. He pulled his arm away. "I'll tell you what, Samantha. I'll give you a week to fire Geo. Otherwise you're gone too. As for this? You and me? It's over."

His shoelaces still untied, Cronk stormed out of the room. She felt bad about hurting the old man. But she knew he wasn't about to give up this adulterous relationship.

He had declared their affair was over twice before, only to beg her to take him back. The first time she had commented that his penis looked like the stubby, brown cigars he smoked. The second she complained that she had to work too hard to get him aroused, insulting him.

Even if Cronk made good on his threat to dismiss her, he would quickly regret it. Sensing she would need leverage, she had secretly videoed this last session, propping up her iPhone on the room's desk, its camera facing the bed.

All she had to do was show Cronk the video and let him know that if she was fired, the next person to view it would be his wife. She tapped the button to end the video session.

"That's for you, Geo. For us." She got dressed and headed into the approaching night.

At the same time Sam was driving home, Geo and Gina were on State Road 84, aiming toward Markham Park to hear Saul Finkelstein's stargazing lecture. They were holding hands and listening to a 1980s tune on the radio.

"What kind of music do you like?" Geo asked.

"Anything with a Latin rhythm. My favorites are Juan Luis Guerra and Laura Pausini."

"Never heard of them."

"Of course, you haven't. You're as gringo as it gets."

He smiled. "Change the station. Find something you like."

"Okay."

While Gina searched for another station, Geo's phone rang. Seeing the caller ID, Geo grimaced. "My pain-in-the-ass brother," he groaned and answered the call. "Hey, Mickey. What's up?"

"Fucking doctors," his older brother snarled, sounding inebriated.

Geo put the call on the Jeep's speaker system, allowing Gina to hear both ends of the conversation. "What's the problem, Mick?"

"If doctors had any compassion, they wouldn't tell you that you're dying. They'd just pull the sheet over your head when everything's said and done."

Geo stiffened. He knew Mickey lived a dangerous lifestyle, drinking heavily, gambling and consorting with loose women. The man had done time in prison. Now he was dying?

"What the hell's going on, Mickey?"

"They should let people have a little hope until the bitter end. But doctors enjoy telling people they're terminal. Damn them all to hell!"

"Are you sick? Dying, for Christ sake? Do I have to come out to California?"

Mickey's breathing was labored. "No," he grumbled. "But I wish I was. They told my girlfriend she's got three months. If she's lucky." He released another expletive.

"I'm sorry. You want me to come out there? Help you take care of her?"

He wasn't sure why he made the offer. He and Mickey were far from close. While Mickey was an irresponsible loser, Geo was trying to be a successful journalist. The disparity generated tension between the two.

"No, thanks, little brother. But you can do me a favor."

"What?"

"Write a story saying doctors are lousy motherfuckers. She didn't need to hear that. Now she has no hope, and she's fucking miserable."

I wish I could. But my editor would prefer I write about giant tomatoes, model trains and knitting circles.

"I'll see what I can do, Mick."

"Yeah. Right," he mumbled and hung up.

"Your brother has a point," Gina said. "People would be much better off if didn't know they were dying."

Geo kept his eyes on the road. "Maybe. But that's never going to happen in this cold, cruel world, where you have to decide whether you want to be buried or cremated."

"I guess."

He changed the topic. "You think your brothers and sisters in Mexico would like to attend our wedding?"

"They don't care about me. Even if they did, they couldn't afford the plane fare. Besides, it's not going to be a big wedding, Geo. It's just you and me in a courthouse."

"We'll need a witness."

"How about your friend Saul? He can bring an extraterrestrial for his date."

Geo grinned. "Good idea."

20

A sizeable crowd attended Saul Finkelstein's lecture at the Markham Park astronomical observatory on Saturday evening. Most were drawn by the title, *Are the aliens already here?* which Finkelstein promoted on several astronomy websites.

Anticipating a good turnout, he held the talk on the observatory's spacious back lawn. With help from the local astronomy club, he set up a portable screen for his presentation and placed fifty folding chairs on the grass. The club provided hot coffee, punch and donuts, all placed on a side table.

Many people brought their own chairs and blankets. Among them were amateur astronomers who enjoyed studying the night skies, and families looking for cheap entertainment. Geo and Gina found a couple of chairs near the front. Both wore sweaters to combat the chilly air.

Using a microphone plugged into an amplifier, Finkelstein began his talk under a clear sky featuring a full moon.

"Good evening, folks. Let's get to it. Even as I speak, aliens from other worlds are probably here on Earth and possibly among us this evening. Oh, you won't recognize them; they know how to blend in. But that guy or gal next to you just might be from a planet billions of miles away."

That drew chuckles. Gina lightly elbowed Geo. "This guy's a hoot," she whispered.

Geo nodded politely.

Finkelstein, a robust man with bristly black hair and a bushy mustache, looked similar to actor Elliot Gould when he starred in *M*A*S*H*, a 1970s hit movie. Despite the brisk temperatures, he wore a short-sleeve Hawaiian shirt, revealing forearms almost as hairy as those on a chimp.

He continued. "I know. That sounds crazy, like stuff that only happens in *Men in Black* or *District 9*. But if you don't think an advanced race can reach earth from billions of miles away, consider this: We humans already have dispatched spacecraft billions of miles into the universe.

"One example: NASA's New Horizons spacecraft flew more than four billion miles into space after being launched in 2006. It took that nuclear-powered ship thirteen years to do so, flying at a speed of thirty-six thousand miles per hour. Now, if we lowly humans can fly way the hell out there, a superior race could do so too, only much farther and faster."

Gina whispered in Geo's ear. "This guy is way out there. Maybe *he's* an alien."

Geo smiled but didn't respond. He wanted to hear the lecture without interruption. Tapping a laptop resting atop a lectern, Finkelstein displayed an illustration of the Earth's Solar System and other solar systems beyond.

"If you don't think aliens exist, consider this: The Kepler Space Telescope, NASA's elite space explorer, has discovered more than twenty-six hundred planets outside our solar system, and many have atmospheres that can sustain life.

"But the universe is a big place, folks, with billions of galaxies and trillions of planets. You can be damn sure that many of those trillions host life forms that make us humans look like baboons by comparison."

Gina whispered in Geo's ear again. "If there are so many aliens out there, why haven't they walked up and introduced themselves? I think this guy is a baboon."

Geo whispered back, "I know it's far-fetched. But let's listen to what he has to say, okay?"

Gina finally got the message. He wanted her to shut up. Finkelstein punched up an illustration of a spacecraft that looked like a satellite dish.

"Now, I'm not the only one who thinks intelligent life is out there. NASA's brain trust does too. They dispatched a space probe, called *Voyager*, to the distant reaches of the universe with an encrypted puzzle onboard. NASA hopes an advanced civilization will decipher that puzzle to learn what we humans are all about. Who knows? Maybe they'll get back to us."

More chuckles.

Finkelstein showed a star with seven planets circling around it. "By using high-powered telescopes on spaceships, scientists have discovered a dwarf star called TRAPPIST-1. It's thirty-nine light years away and has seven planets rotating around it.

"Though small, it generates enough heat to make the atmosphere on some of those planets remarkably similar to that of Earth. Perhaps an advanced race living on one of those planets has arrived on our planet, maybe even on that shuttle out in the Everglades."

Gina sat up straight and stared at the screen.

"Something bothering you?" Geo quietly asked.

"He's full of shit, that's all. If aliens are here, why are they hiding? Why such mystery?"

"Maybe because humans are so dangerous."

"Yeah," she conceded. "That must be it."

Finkelstein displayed a photo of a professional-looking woman with a bright, red-lipstick smile.

"Some of you may recognize this pretty red-head. Her name is Bettina Rodriguez Aguilera, a Miami resident and mother of two children. She was a Republican candidate for Congress a couple years ago. She also claimed that as a seven-year-old child, she was abducted by aliens."

That aroused snickers. Finkelstein chortled with them. "That's right. A respectable candidate, endorsed by the *Miami Herald*, claimed three aliens, including two females and one male, all blond, big-bodied beings, took her onboard their spaceship.

"She distinctly remembered the ship had round seats and was controlled by quartz clocks. Whether they whisked her back to their planet, Ms. Rodriguez Aguilera never made that clear. But she did say that over the years the aliens continued to communicate with her telepathically and taught her few things.

"For instance, they told her the world's energy center is in Africa and that, quote, unquote, 'God is a universal energy.' She made a respectable showing at the polls despite many people thinking she was a crackpot. However, she did lose in the August 2018 primary."

Finkelstein lowered his head for a moment and then looked up at his audience. "I, for one, believe her. I'll tell you what else I believe: the aliens who visited her are probably the same ones who landed in Area Fifty-One many years ago."

"Here we go," Geo murmured.

"I've heard of Area Fifty-One," Gina whispered. "It's in Mexico."

"Nevada, but not that far from Mexico."

"Oh."

"Here's the deal with Area Fifty-One," Finkelstein continued. "In the 1940s and 1950s, there were numerous UFO sightings and even evidence of alien landings in the deserts of Nevada and New Mexico. In 1947, a farmer named William 'Mac' Brazel found metallic rods, pieces of plastic and silvery paper scraps on his land in Roswell, New Mexico.

"He called the authorities, who inspected this debris and, without explanation, carted it away in armored vehicles. That opened the floodgates for conspiracists who believed those parts came from an alien spacecraft that crashed on Brazel's farm. They further believed the government wanted all this hush-hush."

Gina elbowed Geo again. "I'll bet that debris came from that farmer's garbage. It was probably a TV dinner tin container," she speculated a bit too loudly.

A couple of nearby people put index fingers to their lips. "Shhhhh!"

"Shhhhh yourselves," Gina shot back.

Finkelstein displayed a map showing Nellis Air Force Base and its surroundings. "The conspiracy theories were bolstered in 1955, when President Dwight D. Eisenhower approved designating a corner of Nellis Air Force Range as Area Fifty-One. That range, an hour's drive north of Las Vegas, was where new spy aircraft, such as the U2, were tested.

"It was also next door to the Nevada Test Site, where nuclear devices were first tested. So, this was a pretty spooky place where a bunch of covert stuff was going on. The prospect of aliens roaming around in the desert made it even creepier.

"The government initially denied that Area Fifty-One had anything to do with UFOs. But many people didn't buy it. They were convinced the feds were experimenting on aliens and studying their spacecraft in hopes of stealing their technology.

"That, of course, seemed laughable until 1995, when Ray Santilli, a British entrepreneur, released a seventeen-minute video showing an alien autopsy after the creature purportedly crashed near Roswell. Although the video was determined to be a spoof, a lot of people thought it was legit.

"In any case, Area Fifty-One was considered Alien City, even though the government for decades denied that it even existed. Then, in 2017, the Pentagon confirmed to the *Washington Post* that there was, indeed, an Area Fifty-One. The government further admitted to spending twenty-two million dollars annually to investigate UFOs.

"Today, the Department of Defense receives several reports of UFOs every year. Specifically, a little known, shadowy section of the Pentagon, called the Advanced Aerospace Identification Program, investigates sightings by military pilots."

Gina harrumphed. "What a bunch of horseshit. People see a blimp and think it's a UFO."

"Gina! Let's just listen, okay?"

"Sorry."

"For some reason," Finkelstein was saying, "Navy pilots tend to encounter UFOs more often than other military branches. They reported seeing strange objects, appearing like spinning tops, over the East Coast almost daily from the summer of 2014 to the spring of 2015. The pilots said these objects had no visible engine exhaust yet could zip up to thirty-thousand feet at hypersonic speeds.

"At the end of 2018, several Navy pilots spotted an egg-shaped craft hovering over the Pacific Northwest. No one was able to confirm what the thing was. Yet all of this was reported in reputable publications, such as the *New York Times* and the *Washington Post*.

"So, in conclusion, where there's smoke, there's fire. It's obvious that top levels of government think there's intelligent life out there and very well might have visited here." Finkelstein looked directly at Geo. "Again, it's even possible the shuttle in the Everglades was flown in by aliens."

"My ass," Gina mumbled.

Geo shot a look at her. She held up her palms. *Sorry*.

"Thank you for your attention," Finkelstein said. "Those who would like to look through the observatory's telescope are welcome to do so. In addition to the full moon, there's a nice view of Mars this evening."

While most people aimed toward their cars or into the observatory, Geo and Gina remained seated.

Gina shook her head. "What a crock, aliens zipping to Earth like they're taking a trip to the supermarket."

"Well, I enjoyed it," Geo countered, "particularly the parts about Area Fifty-One and the guy who did the fake alien autopsy. I never knew any of that stuff."

"I didn't either. But I do know that Bettina Rodriguez Aguilera, the woman who claims she was abducted by aliens, is nutty as hell."

"Maybe. On the other hand, she had to be level-headed to run for Congress. Why would she lie and risk being ridiculed about such a thing?"

"Maybe she's hungry for attention. Let's get going."

As they got up, Finkelstein ambled over. "Geo, thanks for dropping by."

"Intriguing stuff, Saul. You almost made me a believer."

"Well, when a bunch of extraterrestrials cart you away in their ship and make you a snack for the ride home, then you'll believe me for sure."

"I hope I give them indigestion."

Finkelstein chuckled. Then he eyeballed Gina. "Who's this lovely young lady?"

Geo feared Gina was going to be less than polite, considering she had been so skeptical. But she smiled at Finkelstein.

"Gina," she replied. "I confess, I don't believe aliens have landed on Earth. But I agree with Geo. It was very interesting."

"Thanks," Finkelstein said. "You know, aliens are the first to deny the existence of aliens. Are you an alien yourself, per chance?"

She grinned. "Actually, I am. An illegal alien from Mexico. That is, until Geo and I get hitched."

Finkelstein looked at Geo. "You're getting married?"

Geo nodded, fearing the big astronomer would ask to be invited. "It's just going to be a small, private affair."

"Best of luck to you both. Just for your information, I'm a notary, authorized to conduct weddings, in case you need someone."

No way, Geo thought.

"Thanks, Saul. I'll keep that in mind."

"Cool. If you have any questions for your next story, Geo, don't hesitate to call."

"Count on it."

After they hopped in his Jeep, Gina turned to Geo. "What if I told you I'm a real alien? Like from outer space?"

He threw the Jeep into gear. "I'd first ask if aliens get horny like us humans."

"I'll show you when we get home. What else?"

"Then I'd ask where your spaceship is."

"Maybe one of these days, I'll show you that too."

21

Sam dropped another dumb assignment on Geo on Monday morning, this one about a cat and a Muscovy duck becoming fast friends in a mobile home park. "Lots of photos and video, please," she as usual demanded in an email.

Looking up from his screen, he saw Sam was working at her computer, door closed. Though he wanted to send her an angry note demanding that she yank her head out of her ass, he emailed back, "Okay."

He set up an appointment with the cat's owner, leaned back in his chair and rehashed Finkelstein's lecture. Yeah, right, aliens are living and working among us humans, just like in a sci-fi flick, Geo thought. He had to agree with Gina.

What a bunch of horseshit.

For fun, he searched the internet for stories about people who supposedly made contact with extraterrestrials, such as the woman who ran for congress. Not surprisingly, he found several accounts about people being abducted, taken on board spaceships and probed. He found all of them thoroughly unbelievable.

Then he spotted a story in *New York* magazine, a mainstream publication, saying it's entirely possible that aliens have secretly invaded our planet. The article lent a modicum of support to his story, which suggested aliens might have arrived in the Everglades shuttle.

But his report had been based on Bud Kowalski having no other explanation and Finkelstein's eccentric theories. Bottom line, he was far from convinced that extraterrestrials had been anywhere near Earth.

He was about to end this fishing expedition when he ran across another story, which grabbed his attention. It was about a man named David Huggins, who asserted that he lost his virginity at age seventeen to a female alien. The two went on to produce more than fifty hybrid children.

Huggins chronicled this celestial love story in paintings, which depicted the extraterrestrial as looking fairly human, save for large black eyes. Some of the drawings showed the two coupling, she on top of him, smiling contentedly. To sell this sexy tale, Huggins produced a documentary entitled *Love and Saucers*, available on his website.

Though Geo didn't buy any of it, something about Huggins' story tugged at him, possibly because he was curious. What would a close encounter with an extraterrestrial woman be like? He guessed it would have to be, quite literally, out of this world. On the other hand, Gina was by far the most thrilling woman he had ever been with, and he wasn't sure he could do much better.

Then he wondered: Why had she been so critical of Finkelstein's talk? She seemed to take offense at the mere idea of aliens landing on Earth. He surmised that being on the run from bikers and immigration authorities made it hard for her to entertain such an implausible notion.

Sighing, Geo returned to the real world. Grabbing his keys off his desk, he headed toward the door to do another sappy feature at his scatterbrained editor's behest. Before stepping outside, he looked at Sam, who was still in her office.

She was staring at him, a funny look on her face. Unable to decipher it, he guessed she was pissed off at him again for general reasons. *Whatever*, he thought, and waved goodbye. She gave him a half-hearted wave back.

Little did he know she was envisioning him kneeling between her legs while she was seated at her desk.

Though he hated to admit it, the cat-and-duck story turned out to be cute, two little animals romping in a backyard. Geo interviewed the mobile home owner and his neighbors, all of whom had the same basic take: If a cat and a duck, two species that don't mix, can get along, why can't humans?

Geo thanked the cat owner and jumped in his Jeep, plotting the first paragraph of the story.

It's not often the cat's meow is a quack.

Feeling his stomach grumble, he drove to a McDonald's at the corner of State Road 84 and Hiatus Road. Rather than do the drive-thru, he went inside. He had just taken a bite of a Big Mac when Katrina Tucker sat next to him at his corner table.

She looked foxy as hell in a tight blue skirt and a white tank top. "Hi Geo. Hope I can change your mind."

"About what?" he asked with mouth full.

"Getting together. Enjoying each other one more time. Maybe more."

That hung out there as he slowly chewed and swallowed. He took a sip of Diet Coke. "How did you know I was here?"

"I saw you driving on State Road 84 and followed. What do you say, after you finish your lunch, we head over to my place? I'll make it worth your while."

She leaned over and licked his inner ear, uncaring that four high school boys at a nearby table were intently watching, captivated by Katrina's statuesque figure. Two of them, who had their backs to her, were craning their necks.

Geo pulled his head away from her tongue. "What the hell are you doing, Katrina? Don't you remember? *You broke up with me* because I can't give you a black baby. Please, do us both a favor and leave me alone."

He knew it sounded harsh. But he wasn't interested in playing games now that Gina was in the picture.

Not one to give up easily, Katrina kissed him on the neck. "Yes, I want a black baby. But that's in the future. This is now. I shouldn't have sent that text. We have chemistry, and I don't want to lose that. Even if I end up marrying another guy."

He put down his burger. "You're already planning to cheat on your future husband? That's just not right."

"She sounds like a real whore to me."

Geo and Katrina looked up to see Gina standing by the table. Glancing outside, through picture windows by his table, Geo saw her cycle was parked next to his Jeep on the side of the restaurant.

"Wha'd you say?" Katrina demanded, standing up and staring at Gina. "Best watch your mouth, girl, or I'll bitch slap you so hard your momma gonna feel it."

Not that surprised to see her, Geo guessed that Gina, like Katrina, had tailed him here. He got to his feet, fearing he was going to have to break up a brawl.

Jaw tight, Gina stepped closer to Katrina. "Leave now and don't bother him again."

"Or else, what?"

"Or I hurt you bad."

"Bring it," Katrina snarled and shoved Gina backwards.

"Cat fight!" one of the nearby boys yelled.

Geo whipped around. "Shut up and mind your own business, punk," he snapped.

The kid jumped up with fists clenched. He was big and beefy, probably a linebacker on his school's football team. "Who the fuck you calling a punk, you piece of shit?"

None of that stopped Katrina from taking a swing at Gina. At the same moment, the big high school kid charged at Geo, growling and planning to make a tackle.

Gina ducked and Geo stepped to the side. Because of their proximity, Katrina inadvertently clubbed the kid on the ear just before he plowed into Katrina at the waist. Both spilled to the ground.

"Mother fuck!" the boy blurted, holding his ear.

Katrina pushed the big kid away and jumped up. "This ain't over, bitch," she warned, shaking a finger in Gina's face.

Geo feared she would go after Gina again. Instead, Katrina shot a dirty look at him, pivoted and walked indignantly out of the restaurant. Other patrons and employees silently gawked.

Gina went to the high school kid, still on the floor, and kicked him in the stomach. He recoiled into the fetal position.

"That's for threatening my fiancé. Get out of here before you feel real pain."

He briefly glared at her and nodded. "Leaving right now, lady." He staggered to his feet and looked at his friends, standing near their table. "Let's go, guys."

After they departed the restaurant, Geo sat down, his half-eaten Big Mac, fries and wrapping containers strewn across the table top. Gina sat across from him.

"Thanks, I think," he quietly said.

"What do you mean you *think*? Listen. I just saved you from that bitch. She needs to know you're my man now. Not her play toy."

Geo stared at her. "How much of our conversation did you hear?"

"All of it. That woman doesn't love you, Geo. She just wants to use you."

Maybe it was because he still had feelings for Katrina and was still hurting over their breakup, but everything about Gina showing up here felt wrong. He wondered why she was being so possessive.

Was it because she really wanted to marry him? Or was it because she, not Katrina, was the one using him? In his emotionally charged state, he wasn't sure.

"Do you love me, Gina?"

"Why even ask?"

"Because all the circumstances that brought us together are suspect."

"What do you mean?"

"I mean, how do I know if you really love me or whether you just want to get married to gain citizenship? Maybe that first day at the park you saw an easy mark and played your part to the hilt. Even in bed."

She flared. "How can you say that, after the time we've spent together?"

"You never told me you love me, even after I told you I was in love with you. Even now, when I asked you directly, you can't bring yourself to say it. What do you expect me to think?"

She got up. "I think I've shown you how I feel. But if that's not good enough, maybe I'd better leave."

"What do you mean leave?"

"I mean move out of your house and move along."

He swallowed. He didn't want things to go that far. But he heard himself say, "Suit yourself."

"Then this is goodbye."

She walked outside, mounted the big motorcycle and roared out of the parking lot.

Trying to process what just happened, Geo drove to his office, head full of murk. Somehow, he managed to write the cat-and-duck story that afternoon. But he kept wondering whether his relationship with Gina had been a farce and fearing it was.

When he arrived home that evening, he discovered that she and all her belongings were gone. Suddenly he hated himself.

22

Unable to sleep, he slipped out of bed before dawn, purchased coffee from an all-night convenience store and headed to work. As he sat at his desk, sipping and sulking, he was more crestfallen than tired.

It was Tuesday, the eighteenth of February, the day they were to be married. It would have been a small affair, held in the backroom of a courthouse. That would have been fine by him. He had fallen hard for Gina, despite all the troublesome circumstances.

Now she was in the wind, and he cursed himself again for letting his anger get the best of him, for accusing her of plotting a fake marriage. In retrospect, her passion seemed far too strong to be a charade.

If you weren't such a dumbass, you could have made love to her this morning and taken her out to breakfast, he inwardly scolded himself.

But no. You had to be a first-class jerk.

Outside, first light was seeping over the eastern horizon. The newsroom was dim and empty. He set down his coffee cup, put elbows on the desk and buried his face in his hands.

Ten minutes later he was still drowning in self-pity when Bud Kowalski called his cellphone.

"The astronomy and astrophysicists have finished examining the space shuttle. The results are pretty disappointing."

Geo sat up, trying to pull himself together. "How so?"

"A majority say they can't explain how it arrived in the Everglades but seriously doubt that it came from another planet. Two think it *is* an alien spacecraft, but they're fringe players. Considering we've all but ruled out the Russians and private ventures, we're back at block one. We'll put out a press release later today explaining all this."

"What's your gut tell you, Bud?"

"That something really wonky is going on here. Despite what most of the scientists say, I'm becoming more convinced this thing actually is alien in origin. I know that's crazy talk. But there's no other logical explanation."

"Jesus, Bud. It's *Men in Black* all over again."

"Yeah, I know."

"Can I quote you, saying that extraterrestrials might be at play here? I need to ask."

"I'll probably get fired, but yes. That is, if you couch it by saying we don't have any other reasonable theories and that I'm desperate to find an answer."

"It's a deal."

An hour later, after Sam settled at her desk, Geo walked into her office and relayed what Bud Kowalski told him.

She shook her head. "Dammit, no. I'm not going down that road again, Geo. No more shuttle stories."

"I know, Sam. But we're talking about a panel of top experts saying they can't figure out where the spacecraft originated and a ranking FAA official speculating that it came from outer space because there's no other explanation. That shows what an incredible mystery that thing has become."

"That's all fine and dandy. Let the *New York Times* report it. Where's my cat and duck story?"

He tried to hide his exasperation. "It's done, in the production basket. Sam, we're exclusive with the shuttle for now. I've got Bud Kowalski of the FAA saying maybe, just maybe, extraterrestrials *actually* flew it here. How can we ignore that?"

"Just listen to yourself, Geo! *Extraterrestrials?* Jesus!"

"Sam, that is exactly what Bud said, and he's one of the most credible people I know. I could get a story online this morning long before the feds put out a press release this afternoon."

Her voice rose. "You've already put your employment in jeopardy with that previous alien story. Joe wanted to fire you. But I went to bat for you. If you write anymore nonsense now, he may can both of us."

Geo's tone matched hers. "He wants to get rid of me even though my story gave his tiny, happy-go-lucky newspaper national exposure. What does the guy want me to do, prove there are little green men?"

"Unless you can do just that, I need you to back away from the shuttle altogether. I've got other things for you to do." She took a breath to settle herself. "There's a guy who built a windmill in his backyard to generate energy for his home. His name is John Morgan. I'll email you his phone. I want a feature and photos before day's end."

He stared at her in disbelief. She wanted a fluffy feature about a windmill at the expense of an international story about the possibility of space aliens landing on Earth? *What the hell was wrong with this woman?* But then this was Sam.

"You're the boss," he relented and got up.

"Goddamn right, I am," she grumbled.

He returned to his desk, ready to give up. His work seemed meaningless to him, and now he didn't even have Gina anymore. He stared blankly at his computer, lost. He was completely unaware that Sam was staring at him from her office.

Peeling her eyes away from him, she got up and walked stiffly to the ladies' room. Along the way, she kept her nose in the air and her back straight, an attempt to appear unaffected by the exchange with Geo.

Inside the restroom, she put her palms flat against the sink counter and hung her head, hating herself for disappointing him. While she previously had enjoyed putting Geo in his place, she now realized that she was pushing him away. And that was the last thing she wanted to do.

With each passing day, she grew more attracted to him, her fantasies and desires increasingly dominating her thoughts. It was getting to the point where she was seriously thinking about making a pass. She knew that would be risky. If he turned her down, working together would become awkward if not unbearable.

But what if he accepted? It could lead to a deep and satisfying relationship. They could keep their office romance secret, just as she had done with Joe Cronk. As for Cronk, she would tell him their disgusting affair was over.

For now, she would keep her feelings in check and continue acting like Geo's editor, not a potential lover. When the time was right, however, she would make her move.

Feeling better, having made this resolution, she lifted her hands from the counter, raised her head to study herself in the wall-mirror and gasped. Then she screamed.

For in the mirror, standing right behind her, so close that she could feel its hot breath on her neck, was the scariest, ugliest creature she had even seen.

Although it had a human-like body, save for a long, swishing alligator tail, its skin was murky-green and scaly. It had a dragon face with long, swept-back ears, a prominent snout and a mouth full of pointy teeth. Its large compound eyes were like that of a fly, packed with thousands of photoreceptive units. And it stunk horribly, like a dead, decomposing frog.

Though she couldn't believe the creature was real, Sam was still terrorized. Shrieking even louder, she tried to scramble toward the door. But the thing put two claws on her shoulders and pushed her back, forcing her butt against the sink counter.

"Don't believe aliens exist? Think again," it said in a creepy, synthetic voice.

It extended a long, purple tongue and licked her face from the bottom of her chin to the top of her forehead, slathering her with a gooey, white substance that smelled like vomit. Shutting her eyes tight, Sam released another blood-curdling scream, the sound echoing off the blue-tiled walls of the restroom.

Sharon Qi, the online producer, burst into the ladies' room. "What is it? What's going on?" she yelled.

Sam stopped screaming and opened her eyes. She only saw Sharon. The creature was gone. Its stomach-churning stench had disappeared. Her face was dry, no gooey substance. Chest heaving, Sam looked around. She peeked under the doors of the two stalls. She saw nothing in either.

She looked at Sharon. "There was a ..."

"A what?"

Sam shook her head. She could only figure the stress of the job combined with her adversarial chat with Geo had caused her to hallucinate. She had taken some Nembutal, a strong barbiturate, before coming into work that morning, something she occasionally did with her doctor's blessing. That too could have caused her mind to play tricks.

"I'm just feeling a little queasy." She went to the restroom door and opened it. "Thanks for checking on me Sharon. I'm heading out for a couple hours."

"Where you going?"

"To get a drink."

"While you're feeling queasy?"

"Yep."

Hands shaking, Sam grabbed her purse from her office, walked quickly through the newsroom and went out the front door, making a point to keep her eyes off Geo along the way.

She was about to get in her vehicle when she noticed a motorcycle pulling out of the parking lot. Although the rider was wearing a black helmet, Sam was pretty sure it was a woman.

23

He was driving home from work that evening, still beating himself up over Gina, when he impulsively hit the gas. The Jeep's engine roared and the speedometer shot up to eighty miles per hour, double the limit. He giggled as the white dashed lines on Orange Drive flew by and roadside vegetation became a blur.

At least no other cars were in the vicinity, so he was able to get away with such recklessness – until a cat bounded into his lane twenty yards ahead. The little animal stopped and stared into the headlights with glowing eyes.

"Dammit!" Geo yelled, cutting the wheel and slamming on the brakes. The Jeep careened off the pavement and sliced into a grassy median adjacent to a canal. The front wheels slid to a halt just short of a sharp drop-off into the water. The cat nonchalantly trotted away.

Shaken, Geo rested his forehead against the steering wheel and closed his eyes, darkness filling his head. He didn't care if someone witnessed his stupidity and called the cops. He couldn't see how things could get any worse.

In time, he sat up and tried to accept that he had lost her forever. He backed up onto the roadway, leaving behind deep divots in the grass. Driving slowly the rest of the way, he planned to fix himself a double once he got home.

When he pulled into his townhouse complex, he jammed on the brakes, eyes wide and jaw slack. For there was her motorcycle parked next to his spot. Heart thumping, he parked and dashed inside.

She was in the front foyer stark naked. Without saying a word, she kissed him gently on his lips, neck and cheeks. Taking his hand, she led him to the living room couch. With a subtle smile on her face, while he remained standing, she peeled him out of his clothing.

"Gina, I'm so sorry for – "

"Shut up," she whispered.

Their lovemaking was intense and passionate. For him, it was particularly joyous and thrilling after fearing he had lost her. When it was over, she lay by his side.

"I knew you didn't want me to leave," she said softly.

"How did you know that?"

"We aliens are very good at reading people's minds."

"As in a space alien or an illegal alien from Mexico?"

"Whichever you prefer."

Geo wrapped his arms around her and hugged. His spirits weren't just lifted, they were soaring. "You're right. I'm glad this particular alien came back to me."

"I have to tell you something I should have said a long time ago."

"What's that?"

"I love you. I should have said that sooner. But I was frightened to say it too early."

He held her tighter. "I'm so glad to hear that because I love – "

"I know you do." She broke free of his grip and sat up. "I'm starving. Let's go eat."

He smiled at how she always seemed to be hungry after they had coupled. "Can I get dressed first?"

They drove to a nearby Cuban restaurant and gorged on chicken, rice and plantains, saying little as they ate. Yet both were comfortable in each other's company.

Back home, she put on her nightie, got into bed and dropped off like a rock. He didn't mind. He was too drained to make love again yet too wired to sleep. After joining her in bed, he turned on the television and found a flick, *Men in Black II*.

Having seen it before, he enjoyed the opening scene, in which a small spacecraft crash-lands on Earth delivering an alien. Geo realized this relatively modern movie began in much the same way as the 1950s classic, *It Came from Outer Space*.

As he watched the extraterrestrial in *MIB II* transform into a sexy underwear model and gobble down a ruffian in a park, he deemed all these space-creature flicks to be pure silliness. Then he recalled Bud Kowalski saying the Everglades shuttle might have come from another planet and wondered if that could possibly be true.

No fricking way.

When the movie was over, he turned off the television and wrapped himself around Gina's back. She moaned softly. He smiled, happy to have her home again, in his bed again. Best of all, she said she loved him.

Life is good, he thought before drifting off.

They were sleeping peacefully the next morning when someone started pounding on the front door. Bolting up, throwing on some shorts, Geo charged downstairs to chew out the asshole. He was in no mood to be torn from his nice, warm bed by some door-to-door salesman, particularly when he was hoping to get cozy with Gina again.

When he swung open the door, he was suddenly face to face with no fewer than ten immigration agents. Bob Klug, the senior ICE agent, stood forefront.

"Morning, Geo. Your week is up and you two aren't married. So, she's coming with us."

Flummoxed, Geo blurted the first thing that came to mind. "She's not here at the moment."

Klug smirked as he looked past Geo. Geo whipped around to see Gina standing on the stairway, wearing his bathrobe.

"Not smart to lie to the law, Geo," Klug said, handing him official-looking paper. "This is a warrant for her arrest."

Klug started to go after her, but Geo caught him by the arm.

"Come on, Bob. Give us a break. We were going to get married yesterday but something came up. We'll get it done today. Promise."

The fact was, after Gina's abrupt departure the previous day, he assumed their wedding plans were moot.

Klug pulled his arm free. "Too late, Geo." The agent addressed Gina. "Put on some clothes, sweetheart. We're taking a ride to Krome."

She looked at Geo, who nodded, knowing she didn't have a choice. "Do as he says, Gina."

She went upstairs and five minutes later came down in a sweatshirt, blue jeans and sneakers. She calmly looked at Geo, as if to say, *I'll be okay*, walked over to Klug and turned her back to him.

"Shouldn't have gotten cold feet," Klug smugly said to Geo, as he slapped cuffs on her wrists.

Geo almost blurted "Fuck you." Instead, he silently watched the officers escort Gina to a transport van, its windows reinforced with cage and bars. It was one of a multitude of law enforcement vehicles parked scattershot outside Geo's townhouse with lights flashing, drawing a crowd of onlookers.

As the ICE cars maneuvered out of the complex, Geo stood at his front window, desperately wishing he could help Gina but knowing he couldn't, at least not today. The only thing he could do was follow her to Krome.

He quickly got dressed and drove fifty minutes to the detention center in southwest Miami-Dade County. By the time Geo arrived, Gina was in a processing area being prodded by female inspectors. He found Klug doing paperwork at the front reception desk.

"Bob, can I see her before you put her in lockdown?"

Klug nodded. "Yeah. I suppose. Follow me." He led Geo to an outside, fenced-in holding pen. "Wait here. Someone will bring her out to you. May be a while."

Thirty minutes later two guards escorted Gina to the holding pen. She was shackled in handcuffs and ankle chains and wearing an orange jumpsuit. The guards remained nearby as Gina shuffled to Geo. He took one of her hands and studied her. She almost seemed bored.

"When you get out of here, we're going through with the wedding," he assured. "It's something I want to do."

She smiled in a sad way, knowing marriage was now a long shot. Squeezing his hand, she motioned with her head for the guards to take her away. Geo watched them guide her toward the main building, where she would be placed with several other newly arrived prisoners.

As he trudged to the parking lot, Geo feared that Gina faced great peril in this remote facility, followed by deportation. His only hope to help her would be to retain an immigration attorney, an expensive proposition.

When he got to his Jeep, he kicked the driver's side door as hard as he could, leaving a dent. He hated himself all over again – this time for failing to marry her on time.

24

He arrived in the newsroom shortly after ten o'clock on Wednesday, expecting Sam to yell at him for being late. But she wasn't in her office. Flopping into his chair, he stared out the window by his desk.

The McDonald's drive-thru was bustling. Pickups were streaming into the rodeo grounds, delivering livestock supplies. Even the Beer Barn was busy, patrons seeking a morning buzz. People were getting on with their lives.

His was bleak. What if Gina was ousted from the country? Although he could follow her to Mexico, that would require quitting his job and living in unfamiliar territory. She, meanwhile, would probably return to running drugs to survive.

The only solution, he realized, was to keep her in the states, no matter how much time and money that required. Swiveling to his computer, he decided to see what perils and legal hoops she faced at Krome Detention Center.

He found that Krome was built as a Cold War air-defense base in the 1960s and converted into a refugee processing center in the 1980s. After detainees complained of inhumane conditions, it was upgraded with relatively comfortable dormitory pods in the mid-2000s.

Geo was relieved to see that. But how long would she be held in custody before the government potentially sent her packing?

He found that within two weeks, a judge would determine whether Gina was eligible for an immigration bond, given to those who have grounds for asylum. If granted, she would be released. If not, she faced deportation.

Seeing this, Geo knew it would be wise to hire an immigration attorney as soon as possible. However, if he wanted to remain employed, he first had to do that dumb windmill story. Switching gears, he called the windmill owner and set up an interview at his home.

As was the case with the giant tomato and the model trains, the windmill assignment turned out to be fairly interesting. John Morgan, a retired aeronautical engineer, had erected not one but three windmills on a gentle hill in his backyard. The windmills converted kinetic energy from the wind into electricity.

"I'm able to use appliances, turn on lights and even run the air-conditioning thanks to these things, which, by the way, are properly called wind turbines," Morgan told Geo as the two stood near the machines.

"Your electric bill must be next to nothing."

"It's pretty low. Then again, there are days when there's not much wind and I have to rely on the solar panels on my roof. But they don't produce as much power as the turbines."

"Are you doing all this to keep your power bills low or to help prevent climate change?"

"Both," Morgan replied. "Both solar panels and turbines are already widely used around the world, particularly in Europe. In the Netherlands, for instance, General Electric is experimenting with these gigantic turbines, each of which generates enough electricity to light up a town of twelve-thousand homes."

"Wow."

"A lot of countries over there are in a race to see who can build the biggest and most efficient turbines. Scotland alone has enough wind turbines to provide twice the power that country needs."

"Twice the power?"

"Yes, sir. Renewable energy sources will become much more important in this country too, to reduce carbon emissions and our dependence on fossil fuels. I take pride in helping to save the planet. If I can save money along the way, so much the better."

"What happens when there's no sun and no wind? What do you do then?"

"Plug into the grid. Fortunately, that's not often. I predict that within the next decade most every home and business across the nation will use wind and/or solar energy. We have no choice if we want to prevent the planet from burning up."

"I may have to get a few windmills myself," Geo said, though he knew his townhome association wouldn't allow it.

"Don't forget about solar panels."

"Right."

When he returned to the office, Geo banged out the windmill story by mid-afternoon. Then he searched for an immigration attorney to represent Gina. Though he found a highly recommended woman in Weston, she didn't answer her phone. He left a voice message, asking her to call him as soon as possible.

That evening he had a light dinner of soup and crackers and retreated to his den. He was idly watching some comedy show on television and wondering what Gina was doing at that moment, when his cellphone rang.

The caller ID said unknown. Suspecting it was a phone solicitor working overtime, he angrily answered. "Where do you get off hassling people at night, you slimeball maggot? Why don't you crawl back under your rock and go –"

"Hey, Geo. It's Sam. Your editor."

"Oh my gosh, Sam. I'm sorry. I thought you were someone else."

"Obviously. You sound angry."

"Just dealing with some personal stuff." He wasn't about to mention that his so-called fiancée had him on edge, considering Sam knew nothing about Gina.

"Sorry to hear that," Sam said, slightly slurring her words, sounding a bit drunk.

"I'll live." He was curious as to why she was calling so late. It certainly wasn't to send him to a breaking news scene; Sam didn't believe in breaking news. "How are you doing?"

"Not bad. I'd be better if you came over tonight. I still live with my mom. But she's sound asleep in her room. She won't bother us."

Jumbling his brow, he remained silent for a few seconds. "Are you in some kind of trouble, Sam? Do you need help?"

She giggled. "That's just like you, Geo. Always ready, willing and able to help people. No, you silly goose. I'm not calling because I need help."

"Why then?"

"Because I want you," she said quietly but resolutely. "I thought if you came over, we could have some fun."

Now he was really confused. Was she pranking him or actually on the level? "Sam, you sound like you've had one too many. I'm not sure you know what you're saying."

"I know exactly what I'm saying. I'm saying something I've wanted to say from the moment you walked in the door. I think you're very sexy, Geo. It would be beautiful to be with you."

Geo shook his head, though she couldn't see that. "Sam, you realize you're my boss, right? You haven't forgotten that, have you?"

"No, I haven't. But this isn't sexual harassment, Geo. I'm not forcing you to do anything you don't want to do. I'm just hoping you want to."

"Have you forgotten that just yesterday you said my job was in jeopardy if I didn't let go of the shuttle story?"

"Haven't forgotten that either. But you should know I would never fire you, Geo, not even if Joe Cronk ordered me to. I'd quit first. I enjoy seeing you around the office too much."

"I appreciate that, Sam. I really do, but –"

"There's something else. As I think you know, I've been seeing Joe for some time now. I hate myself for that. But when I started, I honestly liked the guy."

"Don't worry, Sam. I'm not about to tell anyone."

"Thanks, Geo. But I'm letting you know, it's over. I've decided, no more. I'll never sleep with Joe again, even if he threatens to fire me."

"I'm glad to hear that, Sam. You're too young and pretty to be with a man who's so much older and taking advantage."

She sighed. "I know. But now I'm putting it all out there, Geo. I'm asking again. How about coming over? I promise nothing will change between us at work. It will be our secret. No one else has to know."

His first impulse was to tell her he wasn't about to jump in bed with the boss; that wouldn't be right. At the same time, he felt like being brutally honest and telling her that he didn't have much respect for someone who slept her way into her job. But from the tone of her voice, revealing vulnerability, he didn't want to hurt her.

"Sam, I can't. I've got some things I have to do. Why don't you sleep it off and save this conversation for another day."

"For another day? Meaning, maybe someday you'll consider my offer?"

"Maybe someday," he said, knowing that *someday* was code for *never*.

"I hope you mean that, Geo. I'll see you tomorrow. And hey, one more thing."

"Yes?"

"I still want you to stay away from that shuttle. Okay? Goodnight."

"Goodnight, Sam." He ended the call and wondered if there was a full moon that night. Things would probably be awkward with Sam the next day, but he had more pressing matters on his mind.

Mainly, Gina. She was in a prison disguised as an immigration detention center, and he feared for her safety. Yet, for the moment, there wasn't a damn thing he could do. He returned to watching mindless television.

Shortly after midnight his cellphone jolted him awake. He was still in the recliner, the television still on. Grabbing the phone off the armrest, he hoped it was the immigration attorney. But it was Bud Kowalski.

"Christ, Bud, it's late.

"Something's happened. Something very weird."

"What's that?"

"Surveillance video shows that a woman entered the spacecraft this evening, though we can't make any kind of identification. She was able to hide her face. She remained inside the ship for a few hours and fled the area on foot."

"Sounds like she was a crew member."

"I guess you could say that. She used a touchpad to open a hatch. I discovered the pad but couldn't figure out how it works."

"What about the guards? Did they see her?"

"That's what's weird. Despite heavy security, she did this so covertly that the military guards never saw her, and they were right next to the ship at all times. It almost seems like the woman was either super-human or, well, not human at all."

Geo was quiet for a moment. "So, you think she was an ... alien?"

Kowalski inhaled. "Let's put it this way. Her appearance is very curious given all the circumstances surrounding the ship, including its vague origins. As I told you earlier, we have to at least consider that possibility."

"Jesus."

"Yeah. I know."

"Bud, I need to write about this."

"I know. But do me a favor. Make it clear that the alien thing is a possibility, not a confirmation."

"Will do."

"Geo, we're calling a presser at nine o'clock tomorrow morning to reveal what we saw on the video. You're exclusive until then."

"Got it. Thanks, Bud."

Geo hung up, ran to his home office and jumped on his computer.

25

He checked his watch. It was one o'clock in the morning. That gave him four hours to write his story, which he planned to post by 5:00 a.m. If he met that self-imposed deadline, he would beat most media outlets to the punch by hours.

All *West Broward Banner* reporters were authorized to place a story online after hours without editor approval if it was important enough. That didn't happen often at the *Banner*, where soft features were the rule. But circumstances certainly called for it now.

He organized his thoughts. He would lead off with officials believing that a woman, seen on video entering the shuttle, was an extraterrestrial. They had no other explanation, considering she easily circumvented military security and opened a hatch, something that technicians had failed to do.

He would unravel the story from there, noting that the shuttle posed an astonishing mystery. He would chronicle previous developments, including the strange dust and how some top experts had speculated the spacecraft was not of this Earth.

Once published, the story would send shock waves around the globe, particularly if large media outlets, such as the *Associated Press* and the *New York Times,* picked it up.

He only hoped that Sam wouldn't read the story until later in the morning, after it had reached online readers. Despite her inebriated overtures, she would no doubt take it down the moment she saw it.

For now, getting the damn thing written was top priority. He squared himself at the computer and began clacking away in the wee hours of night. A small desk lamp near his laptop was the only light in the room.

As he wrote, he couldn't help but gloat. Other media had chided him for suggesting that aliens might have flown the shuttle here. Now he would make his journalistic colleagues look like monkeys – even though he, himself, wasn't convinced the spacecraft came from another planet.

After an hour of pounding out the story, he took a break, closed his eyes and pinched his nose. His thoughts drifted to Gina, how she must have been horrified spending her first night in detention. He was distressed too, fearing she might be assaulted by other inmates.

He again resolved to contact that immigration attorney the moment her office opened for business. For now, he had to complete this story. He reopened his eyes, regained his train of thought and continued writing:

Aside from the mystery woman seen on video, authorities say a combination of factors, including its impregnable nature, its sophisticated design and a layer of dust from deep outer space, suggest extraterrestrials, not Earthlings, built the shuttle.

When he was almost done, he took another break. His eyes ached, his brain was mush and all he wanted to do was sleep. But he still had to come up with a strong ending. He closed his eyes again and tried to think.

He was jolted awake when he heard the floor creak. Sensing a presence in the room, he remained perfectly still. Then he saw a shadow. He whipped around, expecting to see a gunman in a ski mask.

Instead, it was Gina, leaning down to read the story over his shoulder. Had she been a burglar, he wouldn't have been more startled.

"What … what are you doing here?" he stammered. "I thought you were locked up."

"This is my home. Our home. That's what I'm doing here."

She was wearing her street clothing, jeans and a blouse, rather than prison garb, indicating she had been released. But how? Who had posted bond for her in the middle of the night?

"How did you get out?"

"I mesmerized a few people, grabbed my clothes and walked out. I see you're finally writing about my spacecraft. Know who the person in video was? Me. I slept in the ship for a few hours after you kicked me out."

He licked his lips, trying to process whether any of that could be true. Common sense told him it was just Gina being whacko.

"Seriously, how did you get out? Did they release you because they finally believe we're going to get married?"

"Seriously, I used my powers to mentally stun the guards. As for the wedding, sorry, Geo. We're going to have to put it off for a while."

"Why is that?"

"We're going to take a nice, long ride into space together. It will be a while before we return to Earth."

He nodded, thinking she had gone off the deep end. "I see. I guess I'd better pack appropriately, then. I hear it gets pretty cold up there on Mars."

She grinned. "You don't believe I'm an alien from outer space, do you." It wasn't a question.

"Yes, sure I do. Let me finish this story and you can tell me all about it."

He turned back to his computer screen. Despite his fatigue and Gina's distraction, he wanted to get the story online. He began to write the last paragraph when a powerful force took hold of his body. It felt like an electrical current, turning all his muscles into jelly. He literally couldn't lift a finger. He tried to say something but passed out.

"I don't think so," Gina said quietly.

When he came to, he was strapped into the passenger seat of his Jeep. Gina was behind the wheel, foot to the floor, speeding along a dark highway, aiming into the dead of night. The windows were open, cool air swirling around the interior, her long hair fluttering about her face.

"What did you do to me?" he rasped, some movement returning to his muscles.

"I mesmerized you. Just like the guards at the detention center. Only I didn't zap you quite as hard. Believe me now?"

He guessed they were on US 27 based on the flat vegetation and large power poles along the roadway. He turned his head slightly to look at her. “You must have tasered me. Why would you do that?”

“I didn’t taser you, but it’s similar. Trust me, it was for your own good. I’m not sure you would have gone willingly.”

“Gone where?”

“You’ll see.”

When they came to an intersection, Geo saw a sign that said Tamiami Trail, indicating they were in southwest Miami-Dade County. Gina turned right onto the two-lane highway and punched it. Where was she headed? Geo was too out of it to make an intelligent guess.

She drove five miles before turning into a dim, empty recreation area, which had a sizeable parking area, a boat ramp and only one security light. Gina parked and killed the engine.

“What are we doing here?” Geo stammered. “This where you’re going to kill me?”

She laughed. “Hardly. I need you too much to do that, Geo.”

Just then, the night lit up. Gina glanced in her rearview mirror and saw more than a dozen bright headlights pouring into the recreation area. The thunder of powerful engines filled the air.

“Right on time,” Gina muttered.

Geo strained to look out his window and saw the parade of juiced-up motorcycles surround the Jeep. He immediately recognized the bikers as being the Vicious Varmints.

“What the hell?” he rasped.

“Just some unfinished business.” Gina got out of the Jeep, closed the door and looked back through the open window. “Stay here, Geo.”

Walking in front of the vehicle, she leaned against the grill. Buzz, the musclebound biker who had been in pursuit of Gina for stealing his cycle, dismounted from a new set of wheels. He made a “cut it” motion with a hand near his neck, instructing the rest of the bikers to kill their engines.

Along with two other bikers, Buzz moseyed over to Gina. He lit up a cigarette and took a deep drag. "You're actually here. Didn't think you'd be brave enough. Or maybe that should be stupid enough."

"True to my word," Gina said calmly.

"You really gonna do this?"

"I made a deal. I'll stick to it."

"What deal is that?" Geo asked, staggering out of his door.

Buzz laughed. "Aww, ain't that cute. You brought along your boyfriend to see the show."

"What deal?" Geo said again, only louder.

Buzz took another drag. "Your girlfriend called me tonight. Told me she's tired of being chased and willing to do *anything* to get us off her back. So, I told her she'd have to do me and all my boys. She told me to meet her here."

Geo glared at Gina. "You actually did that?"

"I did."

"How did you get his phone number?" Geo asked her, looking askance at Buzz.

"Remember I told you I met him at the Harley Davidson thing? That he took me back to his mobile home? Got rough? I made note of his number then."

Buzz chuckled. "Almost fucked you good that day, missy. Then you had to go and steal my bike. So, I'll just fuck you now."

She shrugged. "Here I am."

"Gina, no fricking way!" Geo yelled.

Ignoring him, she asked Buzz, "Ready to roll?"

He took another drag. "If you only knew. Just one question. Why here? Way the hell out in the middle of nowhere?"

"We won't be bothered out here," Gina replied. "We sure as hell couldn't go to a hotel."

"Got that right."

Buzz tossed his cigarette, grabbed her by the shoulders and whipped her around, putting her back to him. He nodded at two other bikers, who grasped her arms. Buzz shoved her head down onto the hood of the Jeep, forcing her to bend over.

Gina offered no resistance as several other bikers got off their cycles and gathered around her, each anticipating a turn.

Stinking of cheap alcohol and smoke, Buzz snickered. "Got you right where I want you, missy. Now you're gonna fuck all of us, maybe two at a time. How's that sound?"

Gina lifted her head off the hood and looked over her shoulder at him. "Like you're a first-class pig."

Buzz laughed and began to unzip his pants. "You're the one who's gonna squeal."

"Wait!" Geo yelled, feeling most of his muscle control return. "Wait just one second."

Buzz smirked. "Got a problem, you dumb little shit?"

"Yes. I want you to know," Geo stammered, "that you're the ugliest son of a bitch this side of the Mississippi. Or even the Rio Grande. I'm talking dog-ass ugly."

"And you're about to be the deadest," Buzz growled, turning away from Gina.

He was about to deliver an uppercut to Geo's midsection but instead went limp. The two bikers released their grip on Gina and slumped over. All remained on their feet.

Now standing straight, Gina began shifting her eyes from one biker to another. To a man, their bodies went slack, their mouths dropped open. One tried to charge at her, only to stop dead in his tracks.

When all the men were subdued, Gina nonchalantly walked around them, giving each another look. The bikers responded by shedding boots, jeans, shirts, jackets and head gear.

Once all were naked, including Buzz, under the power of her stare, they formed a conga line. Each man snuggled against the back of the biker in front of him. They all froze in this awkward position.

"Like I was going to stick to that deal," Gina snarled at Buzz, who was at the front of the line.

She collected all their clothing and threw it into the murky canal that ran parallel to the road. She sauntered over to Geo, who had been watching all of this in zombie-like amazement.

"Time for us to go, Geo. Thanks again for trying to stick up for me."

She helped him back into the Jeep, slid into the driver's seat and sped west on the Tamiami Trail.

"Give me your cellphone," she yelled to be heard over the wind blowing through the open windows.

He handed it to her, and she dialed 911.

"I just saw some naked bikers in the recreation area off Tamiami Trail, five miles west of Krome. They appeared to be having sex with each other. You should check it out," she yelled to a dispatcher.

She handed the phone back to Geo, who, despite his malaise, grinned. "Teach those guys to mess with you."

"Got that right. By the way, you're not going to get good reception where we're going."

"Which was where, again?"

She gave the vehicle more gas. "You'll see."

26

They turned into Dade-Collier Training and Transition Airport at two o'clock in the morning. Under the glare of security lights, two military guards marched to the Jeep, palms up, forcing Gina to stop.

Both approached the driver's side. One, resting his hand atop his sidearm, asked her to state her business. She winked at the two men, and both collapsed to the ground.

"Don't worry," Gina assured Geo. "They'll come to in a couple of hours. Won't remember a thing."

Three more guards charged at them as Gina drove onto the space shuttle's ramp. Gina smiled at them through her open window, and all three tumbled to the asphalt as though shot with tranquilizer darts.

"How do you do that? Hypnosis?" Geo asked.

"Similar but more potent."

Gina parked a safe distance from the shuttle, shut down the engine and tossed the keys to Geo.

"Let's go, sport. You able to move?"

"I think so," he said, putting the keys in a pants pocket.

They walked to the shuttle, Gina holding Geo's upper arm to prevent him from stumbling. While he stood near the nose, she tapped the touchpad on the side of the ship, the same one Bud Kowalski found. A hatch opened behind the cockpit and a ramp extended to the ground.

"You first," she said.

Still feeling tipsy, Geo stared at the ramp. "Gina, I don't know what's going on here. But I need to get home and finish my story."

"Your story is going to have to wait, like everything else in your life, Geo. We have more pressing things."

"Like what?"

"Like flying about five-hundred trillion miles to a very special place."

A hint of smoke from a distant wood fire mingled with the night air. A bullfrog lounging on a lily pad in the nearby canal grunted. Geo finally put it all together.

His eyes opened wide. "You're a real extraterrestrial from another planet, and this is your spaceship."

"Took you long enough, ace."

"Your claim to be an illegal alien from Mexico was all bullshit."

"To be honest, I've never been to Mexico. But I hear Cozumel is nice."

He shook his head, befuddled. "This can't be happening," he murmured.

"Trust me, *señor*. It is. Now, get on board. We have a long trip ahead of us."

"Right," he said quietly.

He guessed that trip was to her planet, where she would no doubt serve him for dinner, accompanied by rosemary potatoes and a salad. The only question was whether he would be deep fried, broiled or baked.

He attempted to run. But his legs turned into lead, and he stopped in his tracks.

"Don't be frightened," she said. "I'm not going to hurt you, and I'm certainly not going to eat you. What are rosemary potatoes?"

He stared at her. "You can read my mind? You can read my thoughts?"

"Yes, I can. Let's go, chief."

It felt like a giant hand pushed him toward the ship and up the ramp. She followed him inside and used an interior touchpad, embedded on a wall, to close the hatch.

Now imprisoned inside an alien ship, he stood still for a moment, his heart beating fast and hard. Although Gina claimed she wouldn't hurt him, he found it hard to believe that. He figured that once the shuttle lifted off, he would never return.

To make matters worse, she had firm control of his actions, if not his mind. Whatever she had planned, he could only pray it wouldn't end badly.

"Relax. Take deep breaths," she instructed.

He sucked in as much air as possible and slowly released it. After three of these cycles, his heart slowed down. "I'm okay now."

"Good. You had me worried."

He looked around. The interior was dim so he couldn't make out much. At first glance, it appeared to be about the size of a typical airliner, only without the rows of seats. It also possessed a clean, pleasant aroma.

Despite the lack of light, he could tell the walls were made of a smooth material, no bolts visible. There was no cockpit in the traditional sense, no switches, dials or levers. There was no partition to separate the front area from the rest of the cabin.

However, two cushioned, round seats were positioned near the front window. Each seat rested atop a thick, short pole, which he guessed was adjustable. In front of the left seat was a pedestal with a thin, round-surface top.

Otherwise, all he could see were two small green lights glowing at the rear of the cabin. He guessed they were associated with a restroom. Where did she sleep? He couldn't see a bunk.

Gina answered his unspoken question. "I sleep in a bed, which is now stowed. It slips in an out of the floor, like a mattress that unfolds from a couch."

"Oh," Geo said quietly, wishing this was all a bad dream.

"I assure you this isn't a dream, Geo. You're going to realize that soon enough."

He nervously licked his lips. "Whatever this is, I'm begging you, Gina, set me free."

"I'm afraid not, sport. Like I said, I need you too much."

"For what, exactly?"

"You'll see." Gina pointed toward the front. "Sit in the right seat, please. You're going to be my co-pilot."

He did as he was told, knowing he didn't have a choice. The seat, which had a seatback and armrests, was remarkably comfortable, soft and rubbery to the touch. She sank into the left seat.

"No seatbelts?" he asked.

"Don't need them. Before any kind of incident, be it turbulence or a meteor strike, harnesses instantly and automatically buckle you in. They're sort of like the airbags in your Jeep."

"What if we crash into another spaceship?"

"An anti-collision program would prevent such a thing. But if it did crash, this ship goes so fast that the occupants would be obliterated, instantly turned into molecules."

"Nice to know," he mumbled. Then he asked, "How fast?"

"You Earthlings have yet to create a number that would represent the miles per hour."

You Earthlings? He never imagined that he would be referred to in that way.

He noticed that her Spanish accent had disappeared and that her English had improved significantly. She had played the illegal Mexican alien thing to a T.

"Where are we going?"

"I'm not taking you back to my planet, as you fear."

"Then where?"

Their two seats were close enough for her to lean over and peck him on the lips. "Don't worry. You're with me."

Thanks for the direct answer, he thought.

"You're welcome. Let's roll."

With a couple of touches on a pad, sitting atop the pedestal in front of her, she guided the shuttle slowly upward. Looking out his side window, he could see the ship was generating a massive amount of thrust, buffeting vegetation hundreds of yards away.

He caught a glimpse of his Jeep. It too was being rocked by the ship's mighty propulsion systems. Putting his hand in his pants pocket, jangling his keys, he guessed it would be the last time he would see his vehicle.

When the shuttle was a hundred feet in the air, Gina directed it slowly forward. Below, in the still of the night, he could see the airport entrance road, Tamiami Trail and the lights of Miami in the distance.

She tapped her pad again and the ship shot into space. The Earth shrank into a marble and disappeared outside his window, lost in a patchwork of planets and stars.

For anyone watching from the ground, it would have looked like a meteor in reverse, a trail of flame rising toward the heavens. In the middle of the night in the Everglades, however, there were no witnesses, save for some startled alligators and wading birds.

He thought the acceleration would bash him backwards into his seat. Instead, it felt similar to an airliner climbing steadily after takeoff. Seeing that he had been catapulted into outer space, he was too overwhelmed to speak or react. Yet a voice inside his head was screaming.

Where are you taking me? What are you going to do with me?

A soothing feminine voice invaded his thoughts.

Relax, Geo. I need to fulfill a special mission, and I want you with me. I want us to be together.

He realized it was Gina's voice, answering his thoughts. She not only could read his mind but speak to him telepathically.

He nodded, acknowledging that he understood her message – and powers. "What sort of mission?"

In time, I will tell you.

As she guided this enormously powerful spacecraft out of the solar system, he noted that she still looked like a hard-ass biker chick, a cool, collected one at that. But in revealing who she really was, an extraterrestrial from another world, she had clearly changed, her manner softer and calmer. Somehow, that didn't surprise him.

Hoping she wouldn't notice, he pulled out his iPhone and secretly positioned it to capture a video of her at the controls. The journalist in him wanted to document this thoroughly mind-blowing event, even though he feared no one back on Earth would see it.

"Your phone won't work up here, Geo," Gina said without looking at him. "The ship's electro-magnetic systems have disabled it. Sorry about that."

"Outstanding."

He placed the phone on the floor by his feet, planning to stash it somewhere later on. He noticed the floor, like the walls, was made of that same composite, which appeared to be white.

Exhausted, mentally spent, he closed his eyes to retreat into his own world, one where he still ate lunch at the park, put up with Sam's silly assignments, drove his Jeep on rural roadways – and didn't know that aliens from outer space actually existed.

27

He awoke with a start, remembering he was in a spaceship rocking away from Earth, his fate hanging in the balance. Still sitting up front, the interior dim and quiet, he looked at Gina.

Good morning, sleepy head. Get up and look around, he heard her say. Yet her lips hadn't moved.

"Morning, Gina. Talk to me normally, okay?"

"Whatever you say, slick."

He stood up and faced the rear of the spacecraft. She tapped the touchpad in front of her, turning on soft, yellow ceiling lights. He studied the interior. If it were an airliner cabin, it would have enough room to hold two hundred seats. But it was empty.

"Where's the galley? How do you cook?"

"Thought you'd never ask."

She tapped the touchpad, and several portals slid open in the floor. Dozens of modular forms, all composed of white composite, popped up and snapped into place. Suddenly the cabin was full of equipment.

"What's all this?"

"These," Gina replied, "are units needed for space travel."

"Units?"

"I'll show you.

They sauntered along an aisle, which sliced between the forms. Gina stopped and pointed at a section of various-sized blocks.

"This unit is the galley with an oven, refrigerator and a dishwasher, or, at least, as they are called in your world. These are much more compact and efficient. It also has a table, where we can sit and eat. You won't see a stove with burners because meals are automatically prepared."

"Wow."

He followed her to a section that had a flat form and some other blocks.

"This is the sleeping quarters with a king-sized bed. The other components, dressers, if you will, are places to store clothes."

"Nice." He noted the bed was made up of the same white composite and covered with blue blankets and pillows.

"Normally I don't use blankets and pillows. But I brought these for your comfort."

"Much appreciate that. Is there a reason everything is white?"

"White is the color of simplicity, and simplicity is the building block of advanced civilization."

"Oh."

Gina continued leading him down the aisle, stopping to explain more areas.

"This unit is the gym, which has machines similar to a treadmill and a stationary bike and other equipment. This unit is a laundry room. And over here is the recreation room, where I can exercise my mind with games similar to your video games."

"Wow and wow."

Gina stopped at the final section along the aisle and pointed at devices similar to computer screens. "This is the library, where I read and correspond."

Geo looked around again. "It's all amazing. But there's something missing. I'm not sure what."

"I know what."

She tapped a touchpad and the walls on both sides of the fuselage slid open along the entire length of the cabin. It was as though someone opened curtains over enormous picture windows. Geo's jaw dropped, for outside was a startling view of space, millions of twinkling lights.

"Wow," Geo whispered.

"You said that already."

"It's the only word that comes to mind." His eyes shifted to Gina, who was enjoying his wonderment. "You have a restroom?"

She pointed at a door in the tail area.

"It's the only unit permanently affixed."

"May I use it?"

"Touch the pad next to the door."

He did so. The door slid open and soft lights blinked on. Once inside, he saw the spacious restroom had a sink fitted with a simple faucet and a shower stall big enough to accommodate a horse. The commode looked remarkably similar to the ones in his home. All of the fixtures were composed of the white composite.

"Even space aliens need nice bathrooms," Gina said, still standing in the aisle. "There are fresh toothbrushes and toothpaste in the cabinet. I brought them from Earth for you."

"Aliens don't use toothbrushes and toothpaste?"

"We have special appliances for oral-hygiene."

"Should have known."

She poked her head inside the restroom and pointed at a pad next to the door. "Tap this to open and close the door."

"Thanks."

After closing the door, he relieved himself, washed his hands and brushed his teeth, the water automatically pouring from the faucet. He found no need for a towel; the droplets on his face quickly dried. When he emerged, she was standing nearby, waiting for him.

He asked, "You need to go?"

"Not yet."

He noted she had stowed all the sections into the floor, the cabin empty again except for the front seats. "You have this thing on autopilot?"

"You could say that. It has systems akin to computers on airliners, only considerably more advanced. Let's return up front."

"Okay."

As he sat in the right seat, he noted the shuttle was flying incredibly smoothly, seemingly stopped dead in space. Yet outside he could see planets, stars and galaxies gliding by in much the same way cities and towns pass under the wings of an airliner.

She retook her seat and tapped the pad in front of her, conducting some function.

He regarded her. "I should have known you weren't really from Mexico, that this spaceship was yours. I should have figured out that you were an extraterrestrial."

"Up here, you're the extraterrestrial. How could you have figured it out?"

"The day we watched that movie, *It Came from Outer Space*, you said it was phony. I should have put two and two together right there."

"Really? That wasn't much of a clue. When I told you that someday I'd show you my spaceship, *that* was a clue."

"Teach me to be so fricking naïve," he said. Then he realized something. They weren't floating around the cabin as astronauts normally did in outer space. "Why aren't we weightless?"

"This ship has a gravity element similar to pressurization on your airliners."

"I'll bet NASA's engineers would love to know how that works."

"Relatively speaking, your planet's space technology is still in the dark ages."

Geo had to chuckle. "Our entire planet is still in the dark ages, considering we're trying so hard to destroy it."

She smiled politely. He had an urge to hold her hand but somehow was afraid. She reached out and took his.

"This hasn't changed, what is between us."

He nodded, relieved to hear that. "You look so human. Nothing like the space creatures in movies."

"I *am* human, one who is millions of years more advanced than those on Earth, however."

"You're human? Were your ancestors originally from Earth?"

"No. The laws of nature apply uniformly across the universe. Life evolves in the same way, no matter the planet, though there are variations. All the alien monsters you see in movies are created by the imagination."

"You're saying humans have been around for millions of years on your planet?"

"And on billions of other planets. Modern humans have been on Earth for about three-hundred thousand years. On my planet, more than five million."

"And in those five million years of evolution, the human body has remained the same?"

"It has evolved dramatically. The outward body has not changed much. Eyes are slightly bigger and the skeletal structure is taller and sturdier. What has really changed are the internal organs, which are much more durable, and the brain. Though only slightly bigger, I am able to use one-hundred percent of it. You are able to use only fifteen percent."

"Aren't I the mental midget," he muttered. "By using all of your brain, you can read my thoughts and speak to me telepathically?"

"Among other things."

"Like what?"

"Control the actions of those with less brain power. That's how I dealt with those bikers and the guards at the airport."

"How close do you have to be to read a person's thoughts or control actions?"

"Fairly close, within twenty yards. That's how far brain waves can travel with any strength."

"Are there thought police on your planet?"

"Thought police?"

"Authorities who can arrest you based on what you're thinking. For instance, if you're thinking about killing someone, could you be arrested?"

She raised both eyebrows, amused. "There aren't any criminals because babies with violent tendencies, inadequate intelligence or deformities are terminated before they're born. There is no compunction about killing the unborn on my planet. So, no, there are no 'thought police.' Freedom to think what you want is cherished."

"Oh." Feeling his stomach growl, he realized he hadn't eaten for almost a day.

She got up. "I'm hungry too. Come join me."

She touched a nearby wall pad, summoning the galley to rise from the floor. She pulled a couple of sub sandwiches out of a refrigeration unit. He noticed they were wrapped in Subway paper.

"Sit," she instructed.

As he sat in one of the round chairs at the galley table, she unwrapped the subs, which were filled with meatballs, cheese and tomato sauce, and put them in an oven, which worked much like a microwave. She also produced two cans of Diet Coke and two small bags of potato chips. In seconds, he had the meal in front of him.

"I bought a bunch of Earth food for the trip. Thought you'd like that."

He took a bite and chewed hungrily. All things considered it was delicious. "Thank you. That was very kind of you."

"Once we've consumed the Earth food, you'll have to settle for my normal fare."

"Which is?"

"Not in any form you're used to. So, enjoy while you can."

As they quietly ate, he had so many questions that he didn't know where to start. They came out randomly.

"Where are you from?"

"A planet called Mingor."

"Where is Mingor?"

"Forty light years from Earth."

"But you say that's not where we're going."

"We're going somewhere more distant, about eighty light years away."

"Where is that, again?"

"You'll see."

"How long will it take us get there?"

"About three days in Earth time."

"Three days? How fast are we going?"

She took a bite of her sub. "Let's put it this way. In a split second, we covered the distance between the sun and Neptune and back."

"Neptune? How far is that from Earth?"

"Almost three billion miles."

"Three billion?"

"Two point eight-five billion, to be more specific."

"We're really hauling ass," Geo mused. "What's powering this thing?"

"An element called neutronium, which is not found on Earth. It's billions of times more potent than uranium, which is used to power your nuclear plants."

He looked out his side window and saw an object whirl by in a blaze of fire and vapor.

"Tell me again. How can we be going so fast yet not collide with any planets or comets?"

"This spacecraft is equipped with an anti-collision system that can tell if an object billions of miles away is in conflict with our path. If so, it will make course corrections well in advance of a potential collision."

He nodded, marveling at the spacecraft's capabilities. "Why did you ride up to me at the park that day?"

"I wanted to meet you."

"Why?"

"I did my homework before I left Mingor. I identified you as the man I would like to connect with."

"Why?"

She took a last bite of her sandwich. "This will become apparent as time goes on."

He wanted to press further but sensed she was keeping the true mission of this space odyssey a secret. "Why did you pretend to be from Mexico?"

"Because I naturally look Hispanic, as does everyone else on my planet."

"Everyone looks Hispanic?"

"In terms of skin color and basic facial features, yes. That's the result of millions of years of interracial breeding. All races have blended together. The same thing will happen on Earth, if it survives long enough."

Katrina would be disappointed to hear that, he thought; she insisted on having a black baby.

"How did you come up with that whole story about coming from Mexico, about your name being Gina Martinez, about being forced to work for the cartel?"

"I didn't make it up. There actually is a woman named Gina Martinez who lives that life. I copied it for a cover story."

"How did you learn about her?"

"I have access to the life story of every human on Earth through our intel systems. That's how I know all about you."

"You know all about me?"

"Yes. From the day you were born."

"What was the name of my dog when I was a kid?"

"Pepper, a pedigreed collie. Mrs. Hoekstra was your first-grade teacher. The first girl you kissed was Cindy Goyette at an eighth-grade party. Well, actually, she kissed you. Should I go on?"

He shook his head, having no doubt she knew every detail. "What's your real name?"

"Let's leave it at Gina for now."

He realized he still hadn't asked the most important question.

"What's going to happen to me?"

"Nothing bad. For now, that's all you need to know."

Wishing he could believe that, he nodded. "Okay."

She got up, tapped a touchpad and flipped off the ceiling lights. The interior went dim again. Returning to him, she extended a hand. He took it and she pulled him to his feet.

"Let's make love," she said.

28

Her kiss was wet and eager as their bodies writhed and wrestled in the cushy bed. He was lost in breathtaking bliss, all his senses heightened to a stellar level.

This time, when he experienced her electric thrill, he knew the electricity was real, generated by impulses from her powerful brain. She was hitting him with a mild current to envelop him in ecstasy.

At the moment of release, his body convulsed as though hit by a stun gun. She smiled, seeing his pleasure, and collapsed on top of him.

"Enjoy that?" she whispered as they gasped for air.

"You don't even need to ask," he responded, tingling and astounded that he had been intimate with an extraterrestrial in deep outer space, no less. "I've never felt anything like that in my entire life."

"That was the goal."

He was quiet a moment. "Have you been with a lot of men on your planet?"

"Enough to know they have devolved, not evolved, when it comes to pleasing a woman."

"Really. Why is that?"

She nuzzled her face against his chest. "On my planet, sex is primarily to procreate the species, not make love. True love and passion are rare."

"Is there a reason for that?"

"It's complicated. Let's leave it at that for now.

"Okay."

Although apprehensive about what lay ahead, he felt secure being this close to her. He soon drifted off, his arms around her. When he awoke, he rolled over to touch Gina but only found her pillow. Sitting up, he didn't see her anywhere in the dimness. Had she gone for a spacewalk?

Then she emerged from the restroom and sat on the bed. Her hair was damp and she smelled of pleasantly sweet soap after showering. She was clad in a dark-blue jumpsuit and white sneakers.

"Have a good rest?" she asked.

"Wonderful. How long was I out?"

"More than ten hours."

"Ten hours? Christ. I haven't slept more than six hours in my entire life."

"Get used to it. Outer space is like a rainy day on Earth; you tend to sleep better."

"What time is it?"

"Two-thirty in the afternoon, Eastern Standard Time. It's Thursday, February twenty. But that's on Earth, and Earth time has no relevance where you are now."

"Where am I now?"

"With me."

He smelled coffee. *Coffee on an alien spaceship?*

"Get dressed," she said. "I'm making breakfast."

"Sounds good."

She tapped a pad near the bed, and the ceiling illuminated into orange-yellow, similar to sunrise on Earth.

"The light will help you wake up."

He squinted, even though the light was mellow. "Thanks, I think."

While she prepared breakfast, he found his pile of clothes on the floor next to the bed. He dressed, visited the restroom and sat at the galley table.

She set down two plates, each loaded with scrambled eggs, hashbrowns, link sausage and wheat toast, followed by coffee and orange juice. It looked remarkably similar to the breakfast the two had enjoyed at the Hitching Post. She also provided silverware and cloth napkins.

"How did you do this?" he asked, taking a bite of toast.

"I stocked the refrigeration units with all the meals you like, including those from restaurants. The plates and silverware also come from Earth, to answer your question."

There you go, reading my mind again.

She picked up her fork and shoveled some eggs in her mouth. "I like reading your mind."

He looked out the panoramic windows at a mat of deep black interspersed with dots of light. He noted the interior of the spacecraft was consistently warm and comfortable.

"What's the temperature out there?" he asked, also digging into his eggs.

"Four-hundred fifty degrees below zero."

"Brisk."

She smiled. "In the vacuum of space, there are no molecules available to produce energy or heat. So, yes. It's very brisk out there. More like deadly cold."

He looked down at his wrinkled shirt. "I feel sort of messy without clean clothes."

"I packed some of your pants, shirts and underwear. They're in the dressers in the sleeping quarters."

"Thanks. You mind if I take a shower after breakfast?"

Using a knife and fork, she cut a piece of sausage and inserted it in her mouth. "You would be doing us both a favor. Once you're in the stall, the water will turn on at precisely the temperature you like."

"Precisely?"

"Sensors in the shower analyze what temperature your body most desires."

He sipped and savored his coffee. It was strong and black. "And I remember when I used to turn hot and cold knobs. About two days ago."

She grinned. "When you're done with your shower, the water will turn off automatically. Remain in the stall for a moment."

"Why?"

"You'll see."

When they were done with breakfast, he helped her clear the table. She placed the plates, glasses, cups and silverware into a dishwasher not much bigger than a standard shipping carton.

"This ship must have a huge water tank to run a dishwasher and provide water for the bathroom," he remarked.

"It has a relatively small tank, only big enough to provide for our daily needs. But it is refilled every day."

"How does that work?"

"The ship is equipped with a water-generating system. It produces water from hydrogen and oxygen, two elements that can be more space-efficiently stored than water itself. Also, much of the water is recycled."

"Isn't it unusual for a spaceship to use water for things like showers and the toilet?"

"Only for spaceships with zero gravity. Astronauts on the International Space Station use liquid soap and rinse-less shampoo to bathe. Don't ask how they relieve themselves because it's messy and disgusting."

Geo chuckled. "I'm glad this ship is so cutting-edge. I like water. Lots of water."

"Then go take your shower. I'll leave some fresh clothes on the bathroom counter."

"Thanks."

Once he was in the stall, water streamed out of numerous ports, the temperature hot, exactly how he liked it. A few seconds later a sweet, sudsy soap also poured from the ceiling. He giggled at the convenience.

He rinsed, and the water shut off. He was about to step out when he remembered Gina told him to stay put. He heard a quiet hum and felt dry heat envelop him. Within seconds he was completely dry, down to the heels of his feet.

He got dressed in the clothing Gina had placed on the counter, jeans, a casual long-sleeve shirt, socks and athletic shoes. He joined her up front.

"That's pretty neat, that carwash-like dryer."

"A dehumidifying system dries the entire bathroom, not just the shower. That's why you don't need a towel at the sink."

He remembered how his face had quickly dried earlier. "Impressive."

She began studying three screens, now illuminated in front of her seat. The screens displayed lines, lights and images the likes of which he had never seen before.

"What are those screens for?"

"Navigation. They provide our exact position and our projected path."

"You seem pretty adept at flying this thing. Are you a certified spaceship pilot?"

"On my planet, many are taught how to maneuver a spacecraft in much the same way people on Earth learn to drive cars. Yes, I'm a skilled pilot."

"Who owns this ship?"

"The community."

"Ah, of course. The community. Whatever that is."

She glanced at him. "It's what it sounds like."

"Do you have cars on your planet?"

"Something similar, small vehicles that can fly."

"Like the Jetsons," he said, guessing Gina didn't know about the futuristic cartoon family.

"Very much like the vehicles on the Jetsons."

"Do you fly these vehicles to the grocery store to get milk?"

"We fly them to different parts of the planet. Our main mode of transportation is our feet because it's easy to walk to outlets to procure necessities. And the only milk we drink is from our mothers after birth."

He nodded. "Tell me more about Mingor."

She gazed at him for a moment. Then she put her face in her hands and began to sob, tears flowing, shoulders heaving.

What the hell did I say?

29

Unsure what to do, he put a hand on her back and waited for the storm to subside. When she finally settled down, she displayed a diagram on a screen in front of her seat. He was unable to decipher it.

She sniffled. "This map gives perspective on Mingor's location. Each dot is a solar system. Mingor is three solar system's away from your solar system."

He could only guess that the distance between her solar system and his was significant. "You told me that Mingor is forty light years away from Earth."

"It's in the constellation of Aquarius and in the solar system your astronomers have identified as TRAPPIST-1."

He recalled that Saul Finkelstein had speculated the shuttle came from that solar system. The guy was right all along, Geo thought.

"I've heard of TRAPPIST-1. It has seven planets in tight orbits around a dwarf star, right?"

"That's right. Three of the seven have atmospheres similar to that of Earth. Mingor, the third planet from the main star, has an atmosphere exactly like that of Earth. Your astronomers have named our planet, TRAPPIST-1e."

"I'll be damned," Geo said, recalling Finkelstein had noted that planet might have intelligent life. "Do other planets near Mingor have life on them?"

"Yes, but still in early stages. One has marine life and the other, farther along, has primates not yet evolved into humans."

"Damn. So, Mingor looks like Earth?"

"They are startlingly similar. Mingor has vast oceans, mountain ranges, deserts, jungles, marshlands, plains and arctic regions. It has seasons like those on Earth, summer, winter, spring and fall."

"Sounds like I could vacation there."

Gina, eyes still misty, managed a smile. "In terms of human development and culture, it is vastly different than Earth. We don't have cities but rather, as I mentioned, communities."

"Like the one that owns this ship."

"Correct."

"Tell me about these communities."

"They're small. People live in close vicinity, in an area not much bigger than a city block. They are composed of living units and what you would call businesses. Mostly, they have lots of green grass and trees."

"How are the communities governed? Do they have boards, commissions, legislatures?"

"Nothing like that. Rather than formal government, each community has a leader."

"Is he elected?"

"This person is *selected* based on the will of the majority. If I had to compare our way of life to that on Earth, I would say it is similar to socialism, where all in a community donate to the greater good."

"Meaning no one is rich and no one is poor?"

"Yes. But we live comfortably. We desire for little more than what we have. You see, when a community enjoys peace and harmony, you don't need wealth. Our communities are, on the whole, very peaceful."

"Aren't you lucky. On Earth, our big cities are racked with crime, violence, poverty, hunger, racial strife and people living homeless."

"We have none of these problems. Our accomplishments, such as space travel, are complicated. Yet the foundation of our existence is very simple."

"Mingor sounds like it's a pretty nice place."

"It is a beautiful place."

"But you say we're not going there."

"I wish we could. But we can't."

"And that's why you cried."

"Yes."

"Why can't you go back?"

"Because we must go somewhere more important."

"Where?"

"You'll see."

He nodded, again realizing that she was keeping the true mission of this trip close to the vest.

Just then, she sat forward to study a screen, concern spreading across her face.

"What's the problem?"

"Another spacecraft. One that is not friendly. Pirates, actually."

"Pirates? What kind of pirates? Like *arrr* pirates?"

"The kind of pirates who would take us hostage if they catch us."

"Are they coming after us?"

"Yes. But we'll lose them."

Her fingers danced across the touchpad and the shuttle veered sharply to the right. Both she and Geo would have rolled out of their seats had not the harnesses instantly deployed from their seatbacks, enveloping their midsections.

She tapped the pad again, and he felt the ship decelerate. He glanced up and saw they were aiming toward a gray planet. Though it initially appeared small, it grew large in his window within seconds, thanks to the shuttle's unfathomable speed.

"Ah, Gina, it looks like we're on a collision course with that – "

"That planet is going to save us."

As they drew closer, he could see the planet looked similar to the Earth's moon, a desolate surface pocked with craters. She continued to slow the shuttle and began descending into its atmosphere.

He thought she was going to land the spacecraft in a deep valley. But she swerved to the right again and zipped around to the planet's dark side. There, she brought the ship to a hover near the surface.

"What are we doing?" Geo asked nervously.

"We're going to hang out here for a while and hope the pirates move along."

"Do you think they saw us?"

"Yes. But they're flying a big, awkward spaceship that can't maneuver as well as this one. They won't waste much time searching for us."

"I hope you're right."

Through the windows along the fuselage, Geo and Gina could see the pirate ship slowly glide into the sunlight side of the gray planet, appearing as an odd-shaped shadow. It momentarily stopped, sending Geo's heart into high gear.

"They know we're here, but they won't chase us," Gina said. "They don't want to maneuver in this darkness."

"You sure?"

"Well, pretty sure."

"Great."

The two kept an eye on the pirate ship for several tension-filled minutes. Gina finally looked up from her screen.

"They're moving away."

"Thank the Lord."

She put the shuttle back on course and accelerated to cruise speed. The harnesses retracted into the seatbacks.

"Your friend Saul was right. Many planets sustain life, including sub-human, human and super-human, as is the case with Mingor."

Still wide-eyed, he wondered how she could be so calm after almost being attacked by space pirates. His fingers were still digging into his armrests. "Is that so."

"Yes. There are quadrillions of planets in the universe, and about ten percent have intelligent life. That means trillions of planets have living, breathing, thinking creatures."

"Trillions?"

She nodded. "Billions more at one time supported human life but were decimated because the humans either killed themselves off or destroyed all their resources. Humans have a tendency to do that."

"I sure hope humans don't go that far on Earth."

"Unfortunately, they're on a destruction course. You can relax, Geo. The pirates are no longer a threat."

"If you say so."

Geo exhaled and looked out his window at the vastness of space. He realized Gina was providing information that no one back on Earth knew and probably could never imagine.

He asked, "If a planet dies, do its people try to find another planet?"

"Usually, people and all other animals die too. The good news is, life starts anew on another planet. This is God's plan."

He found it interesting that an alien with superior intelligence believed in a supreme being. He didn't. His logic told him that if there was a God, there wouldn't be so much suffering and destruction on Earth.

"It sounds like you have a strong faith in God. How are you so sure that God does so much planning?"

She glanced at him. "I am sure of it, and by the end of this journey, you will be too."

30

Suddenly the spacecraft began rocking violently as though ramming into severe turbulence. The rubbery harnesses instantly deployed, strapping Geo and Gina into their seats.

"What the hell is that?" Geo shrieked.

"Just a comet-dust storm."

"Just? It feels like we're being bombed, for Christ sake!"

"It's a common occurrence in space. Now you know why this ship was covered with dust. Relax. We'll get through it."

"Relax? How in the hell am I supposed to – "

She gazed at him. His muscles went slack and his mind eased. It felt like he had been injected with sedatives.

"That's better," he mumbled.

Gina grinned as the ship continued through the storm. The turbulence soon ceased and the harnesses retracted.

She got up from her seat. "Let's get some exercise."

"Exercise? Yeah, sure," Geo said, still feeling as though he had taken Xanax.

He followed her to mid-ship. She tapped a pad on the wall and the floor slid open. Up came the onboard gym she had previously shown him.

She pointed at two modular machines with protruding pedals. "Let's start with stationary bikes to get the blood flowing."

"Not like any bikes I've ever seen."

"They work the same way. Hop on."

As there were no handle bars, he had to sit erect. He began peddling slowly. "This is too easy. Can I switch to a higher gear?"

"The machine will dictate the resistance based on your strength. It will get harder after you've warmed up."

"Should have known."

Gina sat on the machine next to his and began peddling. Ten minutes later Geo was panting while she was churning effortlessly.

"I notice you rely on Jesus Christ when the going gets rough," she said, peddling much faster than he.

Gulping air, Geo shook his head. "I'm Jewish. I don't rely on Jesus Christ for anything except providing an excuse to take a vacation at Christmastime. That's just an expression."

"I thought you're agnostic."

"In light of all that's happening here," he wheezed, "as of now, I am a full-fledged, bona-fide Jew."

She smiled. "I can tell you're not being serious. But to be clear, Jesus Christ doesn't play a role in the universe, at least, not like on Earth."

"That makes sense. Do the people on your planet have different religions, like on Earth?

"No. Mingor has no religion at all. There is no need for faith on Mingor."

"I thought that everyone needs a little faith in something."

"I was referring to the faith people have that God exists. Otherwise, you are right. I have faith in you."

Geo stopped peddling, chest heaving. "You do? How so?"

She also stopped, slid off the bike and tapped a pad on the wall. The galley rose up from the floor, adjacent to the gym. She went to a cabinet and pulled out a bottle of Johnnie Walker Black.

"I have faith that you'll enjoy this."

"You brought scotch?"

"I thought you'd enjoy it."

He arched an eyebrow. "Never thought I'd imbibe in space. Then again, I never thought I'd be in space, period."

"Come, join me."

They sat down at the galley table. She poured the whiskey into a couple of glasses and handed him one.

"What the hell. It's five o'clock somewhere back on Earth."

She held up her glass. "To a successful mission."

He gently tapped his glass against hers. "May your soul rise up to heaven before the devil knows you're dead."

"Whatever."

Both took a sip. He winced as the alcohol bit into his gut. "You have alcohol on Mingor?"

"We have mind-altering substances but not very strong. This is much better. In fact, it makes me very – what's the right word?"

"Tipsy?"

"Horny."

She put down her glass, got up and pulled him to the bed.

Two hours later, as they lolled side by side, he gazed out the ship's windows. He saw profound blackness, no stars, planets or galaxies in sight.

"There's something I haven't seen on this trip. There's nothing out there."

"We've entered a dark zone."

"Which is?"

"An area so far removed from all celestial bodies that there is no available light whatsoever, nothing but pure blackness."

"Sort of like being in the middle of the Atlantic with no land in sight."

"Exactly like that."

It was frightening to think their ship was in the middle of an ocean of darkness, an indicator of how far they had traveled. What would happen if the ship suffered a mechanical failure? They'd be up the proverbial creek, he thought.

"Don't worry," she reassured. "This ship is extremely reliable."

He remembered that his thoughts were an open book. "Me? Worried? Not this tough guy."

She smiled at his sarcasm. "You'll be fine after we get where we're going."

"Which is downtown Fort Lauderdale, right?"

"No."

"Tell me something. How did you really learn to speak English? You told me you were a good student, which obviously wasn't the case."

"I used a primer to convert my means of communication into yours," she answered. "Then I read a dictionary and a grammar book on your internet, both start to finish. It took me less than an hour to process both."

"An hour? I couldn't read the back of a cereal box in an hour."

"Anyone on Mingor could do it, even those of average intelligence. While I was at it, I also learned Spanish, German and Russian. Just for laughs."

"You know how to speak four languages?"

"Four Earth languages. I would have learned Mandarin Chinese too, but I was pressed for time."

He shook his head, amazed. "Say something in German."

"Ich geniebe es mit dir hier zu sein, nackt."

"Which means?"

"I enjoy being here with you, naked."

"Impressive. Say something in your language."

"I can only do it telepathically."

"Go for it."

She sat up. He heard a sound inside his mind that was a cross between the high-pitched warble of a fax machine and the cry of a porpoise.

"That's one language I'll never master."

"Probably best we stick to English."

"Good idea."

Gina rested her head on his shoulder. "At his lecture, Saul Finkelstein said a spaceship was dispatched with an encrypted message in hopes that an alien civilization would decipher it. Remember that?"

"Yes. Saul said the idea was for the aliens to learn what humans are all about."

"We solved the encryption in less than ten seconds."

"What did it say?"

"It said, 'Life is short; smile while you still have teeth.'"

"Really?"

"No. It was a quote from Franklin D. Roosevelt: 'There are many ways of going forward, but only one way of standing still.'"

He chuckled. "That sounds more like the kind of humorless bullshit our scientists would send into space." Then he quietly asked, "Why did you cry before when I asked you about Mingor?"

"I miss it. Just that simple."

"Tell me what you miss."

"The people. For the most part they are kind and gentle. For the most part, we are of a like mind."

"I wish Earthlings were as sensible."

"The reason life on Mingor is so wonderful is because, without exception, all of our leaders are women."

"Really. Why is that?"

"Though men are stronger physiologically, women have much more powerful brains. We can produce electrical waves; men cannot. Nature gave us this function is to protect ourselves and ensure the survival of the species."

"Survival of the species?"

She nodded. "When men rule a planet, as they do Earth, they destroy everything around them. This is because of their innate thirst for power. But with electrical impulses, women hold the power and keep men in their place."

"So, you used your brain power that day at the Hitching Post with the bikers?"

"I didn't even touch them. I compelled them to crumble around that toilet. Same thing when they met us on the highway. I forced them to undress and stack against each other."

Geo smirked. "Good move."

"Though she is not a man, I also had some fun with your editor, Sam."

"How so?"

"After she gave you grief for writing that space aliens may actually exist, I paid her a visit in the ladies' room at your office – as an ugly extraterrestrial from a science fiction movie. I think that got her attention."

He had to laugh. "Serves her right, I guess."

"Something else that might surprise you about Mingor. Men and women do not stay with one partner. They have multiple partners. This is accepted."

"How does that work?"

"It's a human tendency to be attracted to new partners. The average life span on Mingor is more than two-hundred years, and people are sexually active for about one-hundred seventy-five of those, too long to be stuck with one partner."

"People on your planet screw until they're a hundred seventy-five?"

"Usually older."

He thought about that. "How old are you?"

"You shouldn't ask a woman her age."

"I know. But I'm asking."

"Seventy. On my planet, I am considered very young."

"You're seventy? Really?"

"Really."

"Well, smack my ass and call me Sally. You don't look a day over sixty-five."

She lightly rapped his arm. "Silly Earthling."

"Have you had multiple partners?"

"Yes."

"Will I end up being just another one of your conquests?"

"Probably, yes. But we can still be together for a long time."

He sat up and looked at her. "If you know that we're eventually going to part, why have we been making love all over the place? Why in the hell *am I here*?"

She pulled him back down. "Because, as I told you before, I love you. I want to be with you. I'm not thinking of the end, of the day we part."

He wondered how she could say she loved him when she was already planning to leave him. Then he also had to wonder – was she using him for her secret mission?

31

He awoke several hours later, his internal clock telling him it was early morning due to the dim interior and pitch-black outside. Yet he couldn't be sure. His watch, like his phone, had stopped working.

Sitting up, he found himself alone in bed again. Then he saw Gina up front, staring out her side window, wearing the blue jumpsuit. He pulled on his jeans and padded over to her. Chest bare, he circled his arms around her shoulders.

"Good morning."

"Good evening. It's six p.m., Friday, the twenty-first of February. That is, in Florida."

He smiled, impressed that she knew that. "What are you looking at?"

"My home."

Peering out the window he saw a planet that resembled Earth with blue oceans, cloud patterns and green land masses. They were flying so low that it filled much of the window.

"That's Mingor?"

"Yes."

He could tell she had slowed the shuttle to better view her planet. "Why don't we land? I mean, we're in the neighborhood."

"I wish we could."

"Why can't we?"

"I'd rather not talk about it."

Is she avoiding one of her lovers?

"That's not the case," Gina said, reminding him that she could read his thoughts. "All that matters now is that we must go somewhere else."

"Understood." He knew it would be best to let it go.

Gina tapped the pad in front of her seat, prompting the shuttle to accelerate and climb. Geo saw Mingor quickly disappear. He was about to return to the bed.

She caught his hand. "You are my lover now, the only one who matters to me."

"I know, Gina, but – "

"But nothing." She got up. "Come. I've got something you'll like."

"Let me put a shirt on first."

"No. I like you that way."

He shrugged. "Okay."

She led him to the galley and had him sit at the table. Much to his continued amazement, she pulled a pizza box out of an oven unit.

"Awesome," he uttered, opening the box. "An extra-large with pepperoni, and it's hot and gooey."

"Told you, you'd like it."

"How did you manage to –? Never mind."

She produced a chilled bottle of white zinfandel, placed two wine glasses on the table and filled each half way.

She sat next to him and both grabbed slices. He took a lusty bite and closed his eyes in pleasure. Although he had always loved pizza, something about being billions of miles from Earth made this the most delicious pie he had ever tasted.

"Pizza and wine for breakfast and breakfast at six o'clock at night," he said, chewing.

She also took a big bite. "I told you, don't try to follow Earth schedules in space. You'll only get more confused."

"If that's even possible."

She wiped her mouth with a napkin. "Now I'm going to tell you the most important aspect about my culture. This will bring into focus some of the things we've talked about."

"You have free pizza delivery on Mingor?"

She smiled tolerantly. "As I told you, there is no need for religion on Mingor. People don't have faith that God exists; they *know* God exists.

We know exactly who and what God is; we have seen God. We have talked to God."

It was an amazing claim, particularly for a super-smart extraterrestrial. He didn't quite believe her. "Everyone on Mingor is on a first-name basis with God?"

She ignored his skepticism. "Because we know all about God, we also know exactly what happens when we die."

"What happens?"

"I'll try to put it in terms you'll understand. God nurtures life throughout the universe. This is done through energy."

"That much I kind of knew."

"Albert Einstein theorized that energy can neither be created nor destroyed. This isn't altogether true because God can create energy. But Einstein was right in that energy cannot be destroyed. Therefore, when one dies, his or her energy is released into the universe and absorbed by other living things."

"Okay. And this means, what?"

"This means the energy from one life gets transferred into many others. When one set of eyes closes, several others open. This is all part of God's plan, the basic reason why God has such great power."

"Sounds like you're talking about reincarnation."

"To an extent, yes."

"So, say a person dies; will he see light again when he opens his eyes as a baby?"

"The simple answer is yes. But when new eyes open, it is with a new soul. Each soul must develop unto its own."

Geo downed his zinfandel and grabbed the bottle for a refill. He wished all of that were true. If so, he could comfort his brother Mickey, tell him that his dying girlfriend will eventually come back.

"I guess that makes sense."

"At the exact moment one shifts into a new life, it's like waking up prematurely from a dream and wishing you knew how it ended."

"That happens to me quite a bit. Where are you going with all this?"

Gina sipped her wine. "Because we, the people of Mingor, know all about God and what happens at death, we are not afraid to die. We can choose to die whenever we want to in a very painless, comfortable way."

What? She's trying to sell me on euthanasia?

"How does that work?"

"There are a few different methods. The most common one is a drug that puts a person into a deep sleep from which they never awake. This drug is widely available, like aspirin on Earth. No special authority is required."

"What if a person is too far gone to take the drug? Like in a state of dementia?"

"Others will administer it. In this way, we have overcome what people on Earth call 'the human condition,' where people, as they age, are less able to deal with the difficulties of growing old."

"What's this drug called?"

She thought a moment. "The closest translation I can give you in English is *absolution*."

"Absolution? As in being freed from guilt?"

"As in being released from consequences."

He finished his second glass. She certainly had laid a lot on him, revealing the secrets of the universe over a pepperoni pizza.

"Why do you want me to know all this?"

"Because our knowledge about God allows us to have extraordinary peace on our planet. This is something you should teach the people of your planet – and perhaps end all the turmoil."

He was beginning to understand what was going on here. The woman apparently pictured herself as an apostle, trying to teach him the ways of the universe. In turn, she hoped that he would share that knowledge with the people of Earth.

"I wish I could end all of Earth's problems with that information. But I don't think they'd listen."

"They will if you write about it in your newspaper."

He chuckled. "That little newspaper has the circulation of a marching band. I don't think that would be the best way to spread your message."

"You'd be surprised. When a message is important enough, many will listen."

He shrugged. "Maybe."

When they were done with the pizza and wine, she stood up. He started to get up as well, but she pushed him back down, peeled out of her jumpsuit and straddled him.

"Told you, I like you without your shirt," she whispered in his ear.

His breath caught in his throat. "Maybe I'll go shirtless more often."

Later, as they lazed in bed, she wrapped around his back, he asked, "Why did you come to me disguised as a biker chick?"

"Because that's your fantasy, and I wanted to fulfill it."

"That means you knew my fantasy even before you met me that day in the park. You knew who I was."

"That's right. You asked me before why I wanted to meet you. It's because I selected you above all other men on Earth to be my lover."

He rolled over to look directly into her eyes. "Over *all other men*? You selected *me*?"

"Yes. I was able to determine the true nature of every man on Earth. We have the technology to do this. You were exactly what I was looking for."

"Which is what?"

"A nice man."

"I beat all other men on Earth, roughly three billion of them, for your affections because I'm a nice guy? Maybe you got a bridge you'd like to sell me."

"That's the truth."

"Look, Gina. Get serious. Why am I here with you?"

"In time, it will all be made clear to you."

With that, she got out of bed, pulled on her jumpsuit and returned to her front seat.

Shrugging off his frustration at being kept in the dark, he took his time getting dressed and using the restroom. When he sat next to her, she was monitoring her screens. It was still pitch-black outside.

"We're getting closer to our destination," she quietly said. Then concern spread across her face. "Shit."

"What is it?"

"The pirates. They're back. Closing in fast."

32

The gargantuan pirate ship looked something like an ocean liner with stout wings. As it pulled abeam of their spacecraft, Gina made no attempt to run. At this close range, she knew the pirates could fire cataclysmic weapons.

Because the pirate ship had oversized oval windows, each as tall as a ten-story building, Gina and Geo could see hundreds of creatures on several different decks, gawking at them. All possessed ape-like characteristics, hairy bodies, stooped postures and heavily-boned brows. None wore any kind of clothing.

One of them began playing with himself as he looked at Gina, a lewd smile on his simian face. Geo was about to give him the finger when Gina brightened the interior lights and went to the side window. She unzipped her jumpsuit, exposing her breasts.

"What the hell are you doing?" Geo nervously inquired.

"They want to see a woman. As long as they can get a good look, they won't blow up our ship. They want to have their fun."

"Isn't that comforting."

Gina pushed the jumpsuit down over her hips, revealing powder-blue panties, which he had purchased for her. "It gives us the advantage. They're distracted, not thinking straight."

Numerous other ape-pirates rushed to the windows of their ship to soak in this intergalactic peep show. Many, clearly aroused, began to pleasure themselves.

"Now I understand why they call it 'spanking the monkey,'" Geo muttered. "What's our next move?"

Gina began gyrating her hips. "I want you to slide over to the left seat, no sudden moves."

He did as he was told. "Now what?"

"The pad in front of you has a yellow light, top left corner. Tap it gently."

He did so and felt the space ship decelerate. The pirate ship pulled ahead but slowed down as well, until it was again at their side.

"Why are we slowing down?"

"To lure their pilot to a window. I can see him at his controls, and I can tell he wants a better view."

Gina slowly pulled down her panties, which, as she had hoped, cajoled the ship's operator to abandon his station.

"There's a green light on the pad's upper right corner. When I tell you, I want you to tap it three times very gently."

"Got it."

Her panties were mid-thigh when the pirate ship's pilot arrived at the window. The moment he began touching himself, she instructed, "Tap the green light. Three times. Now."

Using his index finger, Geo followed her instruction. The shuttle shot forward, the enormous acceleration slamming him backward in his seat, the automatic harness instantly enveloping him.

Gina wasn't afforded the same protection. But anticipating the burst of speed, she grasped a bulkhead handle to avoid flying into the back of the craft.

"This is fun," Geo groaned under the stress of powerful G-forces.

"Just hang on."

Within seconds the ship ceased accelerating, the G-forces evened out and his harness retracted. She pulled up her jumpsuit and returned up front.

"Move to other seat, please."

Geo was already on his way. "Are they following us?"

She looked at a screen. "No. We have too big a lead. Which is a damn good thing."

"Why?"

"If I hadn't tricked them, they would have sucked this whole ship into their cargo bay and captured us. They would have enslaved me for the rest of my life and killed you in a horrible way."

He swallowed. "I guess space isn't such a fun place."

"There are billions of planets run by barbaric men who battle for power and want to take over the universe. Theirs must be a such a planet."

"Earth is such a planet," Geo uttered. "Why were those pirates so ape-like? If all creatures evolve in the same way throughout universe, as you said, how could they look like monkeys yet be able to fly a ship like that?"

Gina tapped her pad. "They have the intelligence of advanced humans. But the environmental conditions on their planet must have put their bodies in a delayed state. There are anomalies in the evolution of life."

"That explains it, I guess. But the last time humans looked like that on Earth, it was before the Stone Age."

Gina studied one of her screens. "We're not far from our destination. From here, the going gets tough."

"It hasn't been tough enough already?"

She touched her pad and two rubbery harnesses secured both of them tightly. "We need to strap in tight for what's coming."

"What's coming?" Geo asked nervously.

"Ever been on a hairy roller coaster?"

"The Hulk at Universal Studios in Orlando," Geo replied, recalling the ride's brain-rattling forces, prompting Katrina and him to scream their heads off only a few months earlier.

"That's going to seem like a merry-go-round."

"Wonderful."

Just then, the pitch-black outside their windows transformed into a brilliant phosphorescent green atmosphere, and the first jolt hit. It felt like the spaceship had smashed through a brick wall.

"Jesus!" Geo yelped.

"He can't help you now."

The jolts came one after another, shaking the shuttle so violently that Geo felt as though his teeth would be jarred loose. Then the ship started rolling over and over, seemingly at the speed of an airplane propeller.

"Oh my God!" he screamed.

"Hold on," Gina yelled. "This will take a while."

He endured what seemed to be endless pounding, careening forces. His teeth gritted, hands grasping his armrests, he now understood why his harness was rubbery; if it was any more rigid, it would have sliced through his chest.

Meanwhile, intense lightning, the likes of which he had never seen before, burned into the green atmosphere. Each bolt was thick and wide and flashed for several seconds. Shortly after each flash, he could hear deep, profound thunder reverberating outside their spaceship.

He glanced at Gina, whose head was flush against the seatback. Despite this nightmare, she seemed tranquil. She obviously had been through this before.

As abruptly as it started, the chaos ended and the ride smoothed out. The thunder ceased and the radiant green atmosphere softened into lighter shades.

"What happens now?" he asked, his heart in his throat.

"We're almost there. Look outside your window."

He gasped. They were approaching a phosphorescent green star. Brilliant orange geysers flowed from its surface like decorative fountains.

"A green sun producing orange streams of fire," he said in quiet awe.

"That's not fire. It's the raw essence of energy."

"Energy?"

"You'll see."

Gina maneuvered the ship toward the green surface, not far from where the orange geysers were sprouting.

Geo asked, "Won't we get incinerated this close to the surface? It looks like molten lava."

"No. It's safe."

Using the vertical thrusters, she slowly lowered the ship until it gently settled on the surface, creating ripples. It was only then he realized they had landed on water, a giant ocean. He thought the shuttle might rock like a boat. But it remained stable.

"Where are we?"

Rather than answer, Gina opened the hatch and stepped down the ramp into the green water, no spacesuit, no oxygen tank, no protection.

She turned to him. "Follow me."

He tentatively took the ramp to its end and stopped. Though he feared the atmosphere would be broiling hot, it was comfortably cool. Now he could see that she was standing not in the green water, but *on* it.

He figured the water was shallow, covering a rock formation. When he cautiously placed a foot in it, he saw that it was clear and deep – and felt firm as concrete.

Gina took his hand. "Walk with me."

Although he expected to sink at any moment, he took a leap of faith and began walking with her. His shoes splashed with each step, yet he felt no sensation of wetness.

"How is this possible? How can we walk on water without sinking?"

"Because this is a holy place."

"What do you mean, holy?"

She put a finger to her lips to silence him. "You'll see."

I'll see? Why does she keep saying that?

As they proceeded, he saw a distant horizon defined by the green sea and a purple sky. The orange geysers, large and powerful, circled them. A cool wind ruffled their hair. Suddenly he felt an overwhelming sense of calm.

"We are in God's house, now," Gina said. "This is where God resides."

33

R*ight. This is where God lives. I hope he serves us lunch. I could go for a peanut butter and jelly sandwich.*

Still holding his hand, Gina stopped and turned to him. "I wouldn't lie to you about something like this."

He immediately regretted his thoughts, knowing he had no business being skeptical. After taking such a surreal space voyage, he should have known anything was possible. Gina had proven so.

"I know you wouldn't lie. I'm sorry I thought such a thing."

She resumed walking. "The God you are about to meet is not the God people on Earth envision."

"He's not?"

"God is not a he or a she. God is not a person and does not resemble a person. God doesn't meddle in everyday events or dictate the course of history. God is the sole source of energy in the universe."

"Energy? God is energy?"

"God produces energy. Yet God has supreme intelligence and can answer any request. I am here to ask God for forgiveness and to do me a favor."

"Forgiveness for what? What favor?"

She didn't answer. They continued walking on water for several minutes. The entire atmosphere felt so Biblical that Geo half-expected Moses to appear while the *Theme of Exodus* played in the background.

They came to an area where the green water met a shimmering turquoise sea.

Gina stopped and released his hand. "Wait here."

"I'm not going anywhere."

As she walked onto the turquoise water, the wind blew harder and the orange geysers flowed more forcibly. She stopped about twenty yards away and bowed deeply at the waist. He couldn't see anyone in her vicinity.

What's she doing?

Falling to her knees, she clasped her hands together and hung her head, apparently praying. She remained in this position for several minutes before motioning with a hand for Geo to join her.

He did so slowly. With each step, the winds blew harder and the orange geysers gushed harder. His hair fluttered about his face and he had to lean forward to make progress.

When he finally reached Gina, the winds died down and the geysers relaxed into a mist. He looked down and saw what she had been praying to, a small, fragile flower with a dazzling yellow bloom.

"Geo, this is God. God, this is Geo," she whispered, still on her knees.

The introduction seemed comical to Geo. *What am I supposed to do? Extend a hand? Glad to meet you, God!*

The plant sprayed a cool, green mist on him. All of a sudden, he was filled with a sense of grandeur, unbridled joy, soothing calm and unlimited strength. He felt as though he could sprint to the top of Mount Everest, do a jig and dance all the way back down.

He instinctively fell to his knees next to Gina and bowed his head. Out of the corner of his eye he could see her lips moving, but he couldn't hear what she was saying.

How long they remained bowed, he had no idea. He was filled with so much euphoria that he wanted to stay there forever. But Gina took his hand and pulled him up.

"Time to go." She addressed the plant and whispered, "Thank you, God."

"Amen," Geo added.

As they walked away, the winds kicked up and the geysers erupted, both with the same force as before. Geo was inwardly bursting with glee, filled with a warm, happy glow.

Gina pointed at the geysers. "That's new energy being shot into space, being distributed throughout the universe."

Geo recalled her saying that only God could produce energy. He felt like much of it had been injected into his veins.

"This has got to be a dream," he murmured. "None of this is real."

The entire sky, from horizon to horizon, turned angry red. Lightning bolts began flashing, each producing a clap of deafening thunder. The waters below their feet roiled into waves, smashing into both of them. Just as abruptly, calm returned.

"That," Gina said, pushing wet hair off her face, "was God's way of letting you know that all of this is real."

"Didn't have to be so dramatic about it," Geo muttered, dripping.

They walked in silence back to the spacecraft. Once on board, both quickly dried off.

"The dehumidifying system works throughout the ship," Gina explained.

Both sat in their front seats and watched the orange geysers in the distance.

Geo turned to her, his buzz starting to wear off. "I'm confused. God is a flower?"

"No. God is an all-encompassing force of nature, which can only be seen as a flower, at once portraying the fragile beauty and the enormous power of the universe."

"Why a yellow flower? It looked like a daffodil."

"Yellow is the color of joy, and thanks to God there is more joy than darkness in the universe."

"Why did that flower spray green stuff on me?"

"That was a gift from God. You will now live a long and healthy life."

"Not if I drink myself to death," he said quietly. "How was God created? How was the universe created?"

"The explanations are beyond your comprehension."

At that moment, he noticed that she looked absolutely ravishing and wanted her more than ever. She met his eyes and shook her head.

"Now that we have visited God, there's no need for more love-making."

He stared at her, feeling like this was Katrina all over again. "After all the times we've been together, all the beautiful moments, you're calling it quits?"

"Yes. My request has been granted."

"What request? What happened back there?"

"As I told you, I needed to request forgiveness from God, and it was granted."

"Forgiveness? For what?

"I let a child die. My child."

34

"How did that happen?"

She licked her lips. "Some background, okay?"

"Yeah. Sure."

"As I told you, love and passion are rare on Mingor, only cursory sex for the purpose of procreation. This is because both men and women are constantly looking for new partners. As a result, women raise children with the help of other women in their communities. Men are not part of the family unit."

"That's cold."

"Very cold," Gina agreed. "Which is why I rebelled against the conventions of my planet. I wanted something more. I wanted love and passion. I wanted a real partner."

"What did you do?"

"I left Mingor in search of these things. I ended up with a man from another planet, one where men still ruled. At first, I thought I was in love with him. But I came to learn he had a bad heart. He beat me, even after I became pregnant with his baby. When I tried to leave, he punched me in the stomach and killed the child."

"Good Christ."

"When I returned to Mingor, I received no sympathy or understanding. Instead, I was shunned. The community was angered that I abandoned my planet and allowed the death of a child."

"Did they know you were escaping from an abuser?"

"They didn't care. On Mingor, every child belongs not just to the mother but to the community. They were mad because I allowed *their* child to die."

"Damn."

"So, now I'm an outcast. The only possible way I could return is if I present the community with another child. So, I begged God for forgiveness and to let me have your child. God granted both wishes."

It took a moment for that to sink in. "Wait, you mean –?"

"I'm forgiven and I'm pregnant. With your baby."

Suddenly it all came into focus. From day one, she had strategically seduced him. The biker chick façade. The illegal Mexican charade. The salty afternoons at his townhouse. The lusty sessions in the spaceship. All of it was one big setup so that she could get pregnant.

"You asked God to have my child?"

"Yes. I selected you above all other men on Earth to be the father. In addition to being a nice guy, you have robust DNA, similar to super-human DNA."

"That's very flattering. But you used me."

"Yes."

"You lied about loving me."

She didn't answer.

"I'll take that as a yes as well. What's going to happen to the baby? *My* baby?"

"I'll raise it on Mingor. Where it belongs. Where I belong."

"Will I be able to see the baby? I *am* the father. I should be part of its life."

Gina shook her head. "You're going home."

Geo shut his eyes and hung his head. "Boy, can I pick 'em," he mumbled.

"You shouldn't be annoyed. You should be grateful."

He looked up at her. "And why is that, exactly?"

"I introduced you to God. You are the first human from your planet to see God, to know for a fact that God exists. With that knowledge, you can help all the people on Earth."

Geo forced a chuckle. "Like they'll believe me. A space alien hauled me to God's house and, guess what, God is a yellow flower."

"At least you can try."

"Trust me, it's not worth it."

She shrugged, tapped her pad and directed to the shuttle to lift off. Geo got up from his seat and retreated to the bed, his head throbbing, his heart broken – all over again.

The flight back to Earth was uneventful, no dust storms or pirate attacks. He spent most of the trip in the gym while she remained up front. Though she provided him with food – something that resembled human brains but tasted like oatmeal – they ate separately.

When they finally landed at the Dade-Collier Training and Transition Airport, shortly before dawn one morning, Geo was sleeping in the big bed. Gina shook him awake.

"You're home."

He sat up and saw swamp vegetation under gentle pink light outside the windows. "What day is this?"

"Sunday, the twenty-third day of February. We've been gone for almost five days, Earth time."

"Five lost days," he deadpanned.

Having no response, she returned up front. He got dressed, grabbed his cellphone and stuffed it in a back pocket. He decided to leave behind the clothing she had brought for him.

Seeing he was ready to go, Gina opened the hatch and stepped down the ramp before him. He stood in the doorway for a moment, looked around the ship one last time and followed her to terra firma. The airport was dead. The military guards were gone, transferred to another duty.

He turned to her. "I love you, Gina."

She silently nodded. He knew it was her way of saying goodbye. Forever. She returned inside the ship and closed the hatch.

"Have a good trip," he muttered.

Pulling his car keys from his front pocket, he slowly walked back to his Jeep Sahara. Once behind the wheel, he watched the spacecraft lift off, its thrusters buffeting the nearby vegetation.

As it rose straight up about a hundred feet, it suddenly dawned on him that he should video its departure. It would be proof of the great tale he had to tell.

But when he pulled out his iPhone, it failed to turn on. The battery was dead. He watched the ship hover for a moment and, like a bullet, blast into the sky and out of sight.

First, he smiled, somehow finding all this the perfect ending to his odyssey. Then he began laughing, long and hard, until tears flowed from his eyes.

35

Rather than drive home, he stopped at a Denny's on University Drive for breakfast. He was half-heartedly working on two eggs over easy when his phone rang. He had charged it during the drive from the airport, so it was functioning again.

"Where the hell you been?" Bud Kowalski asked.

"Left town for a few days."

"Yeah, well, about the same time you went AWOL, that fricking thing disappeared. Just vanished."

"The shuttle?"

"Yep. No one saw or heard a thing. The military people can't even account for it."

That's because a space alien zapped them.

Geo pretended to be surprised. "Damn. What do you think happened?"

"We can only surmise that it was the Russians all along. Their people must have snuck into the ship and flew it back. Hell, maybe they had someone hiding inside the whole time."

Geo wished he could tell Kowalski that he had just flown in the spacecraft to the deepest reaches of the universe and visited God, all courtesy of a beautiful extraterrestrial. But he needed to share his story with the whole world, not just one man. That's what Gina wanted him to do.

"So, that whole business about it possibly coming from another planet was just a bunch of bunk?"

"Yeah, definitely," Kowalski grumbled. "Wherever that thing came from, the media is going to have a field day ripping the government for losing track of it."

"Don't worry, Bud. I plan to write a responsible story. You won't look bad."

"Thanks, Geo. I doubt the *New York Times* and the *Washington Post* will share your compassion."

"Oh, I don't know. Maybe after they read my story, they'll change their minds."

"We can only hope. Have a good day, my friend," Kowalski said, hanging up.

Geo finished breakfast and drove to his office. As it was a Sunday, the place was empty. He stretched, shook the cobwebs out of his head and began composing the biggest story of his life. He wrote it first-person to ensure readers would pick up all the nuances.

> *I have met God. Not in a philosophical way but for real, in a physical way. God is not in the image of a human but rather an all-encompassing force that manages all the energy in the universe.*
>
> *God uses a magnificent, yellow flower to represent all this power and can be found in shimmering turquoise waters in the middle of a brilliant green sea on a phosphorescent green star 80 light years from Earth.*
>
> *How do I know this? Because a space alien, an alluring female at that, gave me a ride in her spaceship – the same shuttle that mysteriously landed in the Everglades – to visit God.*
>
> *I not only saw this flower of God but felt it. God sprayed me with green mist, which I'm told will ensure I live a long, healthy life.*
>
> *Pretty unbelievable, right? Wish I could have documented all this with photos or video. But my cellphone didn't work in outer space and died when I returned.*
>
> *So, I'm hoping you'll take my word for it. I'm a level-headed guy who has never experienced mental delusions.*
>
> *If you do believe me, here's what you should know: On those planets*

where the inhabitants know God exists, there is total peace, no war, no crime, no fear of dying.

That's because these inhabitants know that as one dies, another is born, that the energy from one is transferred to another. Not the soul, necessarily, but the energy.

You know what these inhabitants are? Human beings. Because life evolves in the same way throughout the universe, humans exist throughout the universe. This, I was taught by the female alien.

If the humans on Earth could come to believe all of this, we could end all the strife on this planet. We could advance to the next level of existence, as others on other planets have done.

Most already know that in its current state, Earth is in peril. We are ripping each other apart and depleting resources. Isn't it time for a positive change?

From there, he provided a detailed chronology of how he met Gina, how she presented herself as an illegal Mexican alien and how she earned his trust, only to swoop him away.

He related that she came from a planet named Mingor, part of the TRAPPIST-1 solar system, and that women were in charge of that world because they had the ability to control men through mental-electrical impulses.

Because it was an interesting detail, he also noted that it was acceptable on Mingor for women and men to have several partners as the average age at death was more than two-hundred years old. He further mentioned that children were raised by their entire community, not in a family unit.

He neglected to say the people of Mingor shunned love and passion. He left this out because he hated the fact that Gina had initially claimed she loved him but, as it would turn out, didn't.

Though he suspected she would never see his story and thus in no way be affected by it, he still neglected to relate how they had made love several times and that she had gotten pregnant, her primary mission all along. Out of respect for her, he didn't want to get that personal.

The first draft done, he went through it again and added more details, such as the pirate attack and violent comet-dust storm in outer space. Once the story was ready to go, he could have posted it on the *West Broward Banner*'s homepage. Because it was Sunday, no editors around, he had the authority to do so.

But he decided, no. The story was too important, too potentially explosive, to slap online. He wanted Sam to approve it and hopefully give it the promotion it so richly deserved. Despite her propensity to ignore important news events, he was confident she would play this story big.

After all, this could well be the most momentous news story in history. If people read it, if people believed it, it might—*just might*—help elevate human existence to the next level, as Gina had suggested.

If Sam were smart, she would ask Joe Cronk, the publisher, to hire a professional marketing firm to publicize the story before running it. The *West Broward Banner* would suddenly become more widely read than the *New York Times, Washington Post, Chicago Tribune* and *USA Today* combined.

All this in mind, he sent the story to Sam's personal directory so that she could edit it. He then sent her an email marked, "Urgent."

"Hi Sam," he wrote. "I'm sorry I missed a few days of work last week without explanation. But when you read the story in your directory, you'll understand why. It's a huge story with enormous ramifications. Geo."

That done, he went home, stripped down to his boxer shorts and fell in bed, thoroughly exhausted. Although it was late afternoon, he slept like the dead straight through the night.

36

"Geo, get in here!" Sam yelled, standing at the door to her office. "Now!"

He had been sitting at his desk, looking out his window at a cloudy Monday morning. He got up and ambled to her office.

It's showtime.

"Good morning, Sam. I take it you read the story."

"Sit down," she growled, all but slamming the door behind him. She went around her desk and fell heavily into her executive chair. "What the hell was that?"

"A story that needs to be told."

"What happened? You went on a binge and decided to write a fairytale to piss me off?"

"It's no fairytale, Sam. All of that actually happened. The woman from another world. The trip into space. And all of that business about God? Absolutely true. Think of the story as a primer to advance human existence to a better place."

"Right," she said quietly, then shouted, "Are you out of your fucking mind? You must be if you think I'd publish that steaming pile of dog shit!"

He sat up straight. "Listen, Sam. For once in your life, do the right thing. Stop worrying about Joe Cronk and run this story, a story that will make you a very famous editor."

"You mean infamous, don't you? I'd go down as the dumbest editor in history for running something like that!"

"If you read the story carefully, you'll see that it's the God's honest truth, so to speak. Now, make your own decisions. Don't listen to Cronk."

"What are you talking about? You don't think I make my own decisions?"

"The only reason you have your job is because you're screwing Joe.

So, yeah, I think that what he says goes. But you need to override him this time, Sam. Run it as big and bold as possible, online and in print. In fact, you'd be wise to hire a marketing company."

"For your information," she snarled, "the only reason you're still working here is because I keep protecting you. I keep telling Joe that you're my best reporter. But if you think you can get away with this bullshit, I can't defend you anymore."

He had hoped this wouldn't be an uphill battle. "Sam, I'm telling you. I met an alien woman who flew me billions of miles away to see God. You know me. You know I don't have mental problems. You know I wouldn't prank you. As unbelievable as it may seem, all of that happened."

She shook her head. "If you think all of that actually happened, I strongly suggest you see a neurologist. I'm killing the story and giving you another assignment. The Girl Scouts are selling a new brand of cookie, a coconut, marshmallow and chocolate thing. Supposed to be really delicious."

His jaw fell open. "You want me to write a story about Girl Scout cookies instead of about how I met an alien who proved to me that God actually exists?"

"That's exactly right, Geo. Of those two stories, readers will only believe one of them. Now, I need you to get rolling on the cookie story, pronto."

He glared at her. "You're killing the story because I turned you down the other night, right?"

She laughed, though it was forced. "This has nothing to do with that silly call, which, I confess, I made while a bit inebriated. I'm killing it because it's pure nonsense."

He was astonished. Pure nonsense? A *steaming pile of dog shit*? She was capriciously dismissing a blockbuster story that could change the course of human history.

He regretted that he hadn't posted the story without permission. At least it would have been read by some people before Sam took it down.

He stood up and started toward the door, knowing had just run into a brick wall – reenforced with stainless steel. All his enthusiasm to tell the world what happened suddenly drained out of him like blood from a knife wound.

"Don't pull anything like this again, Geo," Sam warned. "Not if you want to continue working here."

"Don't worry," he grumbled. "I won't."

As he walked back to his desk, Sam swiveled back and forth in her chair, wondering what actually had forced Geo to miss four days of work. Whatever it was, despite their spat, she was glad he was still employed at the *Banner*.

That almost wasn't the case. After Geo failed to show up for work on Friday, Joe Cronk had demanded that she "can his ass" or else he would make good on his earlier threat to fire both of them.

That was when she showed him the video of their last tryst. "Go ahead, fire us, Joe. After you do, this video goes straight to your wife. By the way, we're over."

Despite winning that battle, she was beginning to lose hope that she and Geo would ever become lovers. She sensed that his heart belonged to another.

37

Geo was sitting at a picnic table in Tree Tops Park, eating a meatball sub, his favorite lunch. It was a sunny Friday in mid-April, two months after he last saw Gina.

He never could have guessed that her lunchtime interruption, right in this park, would lead to an unimaginable space odyssey. But she was gone now, forsaking his love to return to her planet.

At least he had written a story documenting the entire experience. He had done so on his home computer for his own edification, to keep Gina close to his heart. After Sam rejected it, he had no intention of trying to sell the piece to any other media.

Now he was back to the grind. Taking a bite of the sub, Geo thought about how dramatically things had changed in the past two months.

Shortly after he returned from space, Sam quit as the *Banner*'s editor to become a municipal reporter at a small daily newspaper in Lorraine, Ohio. During the staff goodbye party, she had pulled him aside.

"Visit me in Lorraine if you're in the neighborhood," she had offered, then gave him a peck on the lips.

"Maybe I'll do that," he replied, actually contemplating a trip to Ohio.

Then came another big surprise. Following Sam's departure, Geo expected Joe Cronk to immediately fire him. Instead, the old publisher had called Geo at his desk one day.

"Want to be our new editor?"

Geo didn't hesitate to respond. "It would be an honor, Joe. But why the change of heart? Sam made it clear you don't like or respect me."

"For a time, I didn't. But when Sam turned in her resignation, she recommended you take over, saying no one was more qualified. I realized she was right."

Geo doubted it was that simple. He guessed Sam had some sort of leverage over the guy. But he didn't question the promotion any further. Instead, he requested the staff be expanded in order to mix in hard news with all the features. Cronk readily agreed.

Now Geo was in charge of fifteen reporters, photographers and online producers, triple the number of staff members under Sam's regime, and the small newspaper was as strong as ever.

Done with lunch, Geo remained at the picnic table, savoring a mild breeze. He closed his eyes and remembered her soft touch and hungry kiss. He wished she would roar up on a motorcycle again and –

He caught himself. It did no good to fantasize about Gina. For all intents and purposes, their adventure together had been a dream. He could only hope that she and her child – *his* child – were happy somewhere way out there, forty light years away.

Eyes still closed, he thought about all the tasks he had to do that afternoon, edit stories, lay out the next edition –

His eyes shot open. The world around him went dead silent, the buzz of insects, the hum of traffic and even the breeze ceased. The sunny sky darkened and he was overcome with a sense of dread.

He looked up and froze. For there was the enormous pirate spaceship blotting out most of the sky. Under the light of the Earth's atmosphere, he could see that it had a tubular fuselage, save for a flat belly, and a corroded exterior. It glided over him and stopped.

"No. This can't be," Geo whispered.

The silence abruptly turned into screaming. From his position in the open park, he could see people on nearby sidewalks and roadways running for their lives. But no one was safe.

Thousands of portals on the ship's belly opened and an eerie red light poured out of each. The moment the light detected a person, he or she was sucked violently upwards into the monstrous contraption, kicking and squirming.

Cars offered no protection; the red light would laser open a hole in the roof and yank an unsuspecting motorist and any passengers skyward.

Neither did homes or buildings. The very discriminating light would burn through concrete, bricks and wood to whisk victims upwards.

As for those walking or jogging outside, the light plucked them up so quickly that they were pulled out of their shoes.

Geo felt bile rise in his throat, seeing the mass abductions at the hands of hostile aliens. Then a red light panned in his direction. He jumped up and ran, hoping to hide behind some nearby trees.

He didn't make it. The light washed over him and with ruthless force, tugged him upwards. Along the way, he could see thousands of other shrieking people going up with him, arms and legs flailing. He began screaming and kicking, too.

Once inside the pirate ship, the light dumped him on a pile of other people. He landed on a teenage girl who had been riding her bike. Under her was a young mother who had been pushing her baby in a stroller; the baby was now just another body in the pile.

Geo rolled off all the dazed people and got to his feet. Despite his bewilderment, he noticed he was inside an immense bay, big enough to hold an NFL stadium. The entire interior, floor to ceiling, was coated with a black, rubbery material that had the pocked texture of a sponge. The smell was horrid, like an uncleaned elephant cage.

Meanwhile, the red lights delivered a steady stream of bodies, men, women and children, all crashing into each other, flopping like fish emptied out of nets onto a boat deck.

All of a sudden, thousands of ape-like creatures, the same ones he had seen while with Gina, charged into the bay from every direction. They reminded Geo of the early human ancestors depicted in the opening scenes of *2001: A Space Odyssey*, primate faces, loud grunting and awkward loping.

Armed with prods, they began herding the humans into holding pens around the perimeter of the bay. Each pen was about the size of a bedroom and composed of something akin to chicken-wire.

People were stuffed in so tightly that they were forced to stand.

When one pen was full, the creatures would shove people into the next pen down the line. The prisoners screamed and rattled the structure of their cages but to no avail.

Acting on instinct, Geo tried to run, where he had no idea. Before he could take two steps, several of the grotesque apes grabbed him. Though he fought with all his might, they pushed him into a pen and clanged the door shut.

Although packed in with a crowd of howling and terrorized people, Geo was able to see the horrible next step for the human captives.

The beasts pulled several people at a time from a pen and dragged them to a long table with a marble-like surface. After ripping off their clothing, several apes held down the victims while other apes, armed with meat cleavers, hacked them apart.

The body parts, from head to feet, were tossed onto giant grills. The "cooks" then tossed the sizzling human meat haphazardly into a crowd of hungry space apes.

The stronger ones grabbed the weird-smelling fare first in their bare paws and gorged, guts spilling out of their gorilla-like mouths.

The entire scene was so barbaric that Geo began shrieking uncontrollably like everyone else around him. The wailing echoed off the interior walls, exacerbating the victims' torment. All the while, more humans were hauled on board, processed and prepared for a midday meal.

Eventually Geo fell into comatose state, resigned to being hacked apart, grilled and eaten. Somewhere in the back of his mind, he wished the Air Force would blow this alien ship out of the sky with a powerful missile.

He would be killed in the blast. But that would be better than being tossed onto that table, where the whacks of the meat cleavers were becoming as loud as gunshots.

Suddenly it was his turn. Four ape-aliens pulled him out of his pen dragged him toward the butcher table. Snapping out of his trance, he began bucking and kicking.

The creatures ripped off his clothes and tossed him onto the table, now drenched with blood. While the four apes held him down, one at each limb, a particularly ugly animal appeared with a meat cleaver.

With a gleam in its eye, the butcher ape raised the cleaver high above his head. Geo went perfectly still and glared at the thing.

"You damned dirty ape," Geo uttered. "You stink."

Though it didn't understand, the ape became infuriated. It raised the cleaver even higher and mustered all its strength for the downward swing through Geo's neck.

That's when the ape's head exploded like a shotgun blast through a watermelon. When the flurry of blood and brain matter subsided, Geo saw Gina standing behind the now-headless creature, the hairy body of which collapsed to the floor.

She calmly winked at Geo. "Looks like I got here just in time."

She stared at the four apes holding Geo down. One by one, their heads exploded as well. Pulling Geo off the table, Gina instructed, "Stay close to me."

She began walking around the bay and with her glare blasted the brains of each ape-alien that came into view. Seeing that an intruder was breaking up their party, thousands of the creatures attempted to scamper toward tunnel-like exits. But Gina's laser eyes caught them, and all their heads exploded.

She led Geo to a portal in the floor. Looking down, he saw that her shuttle was right there, hovering below the pirate ship, a hatch in its roof open.

Using an emergency chute similar to those on airliners, she and Geo slid down to the shuttle. Once inside, he found himself in familiar surroundings. Gina handed him a blue jumpsuit, just like the one she was wearing.

"What about all the other people back in the pirate ship?" he asked as he put on the jumpsuit.

"They'll be fine. The pirates know they must transport them back to the ground – or answer to me."

"That's good." His heart was still pounding.

"This is all my fault," Gina lamented. "They followed me here, and I didn't realize it until I saw that they had taken you."

"They followed you?"

"They must have been hiding behind Jupiter while I was on my way to Earth."

"Why were you on your way to Earth?"

"To get you. I want you to return to Mingor and help me raise our daughter. She'll be born in fourteen months."

"We're having a daughter? In fourteen months?"

Gina nodded, joy in her eyes. "Pregnancy for advanced humans takes sixteen months, which is why I'm not showing yet. You'll recall I got knocked up a couple of months ago."

"I remember well." He thought about having an alien daughter. "I'll bet she's going to be a beautiful little girl. I only hope she takes after her mother rather than her dumb father."

"Don't worry. She's already been tested for intelligence, and she'll be smarter than me. She'll have your eyes."

"My eyes? Really?"

"Really."

She came close to kiss him, and the moment her lips touched his, his eyes opened.

He was back in the park. All was quiet save for some chirping birds and rustling tree branches. The sun was bright. His sandwich's wrapping papers were right where he left them on the picnic table.

His Jeep was parked a few yards away. He was wearing his work clothes, not the jumpsuit. Other park visitors were grilling hamburgers and tossing around a frisbee on a nearby grassy field. No alien pirate ship darkened the sky.

He put his head in his hands, realizing he had fallen into a deep, midday nap, allowing his imagination to go into overdrive. The lingering images of the ape creatures were so vivid that it made him wonder if the entire encounter with Gina, from her first appearance in the park to their meeting God, was a dream too.

"Jesus. I've got to stop eating meatball subs for lunch."

38

On Saturday morning, he was in his home office staring idly at a blank computer screen. As the new editor of the *West Broward Banner*, he wrote a weekly column to give the paper more personality and explore community issues.

Usually, he had a wealth of subjects to write about. Davie was a diverse town. But on this day, he was drawing a blank. He briefly thought about revisiting how he had rocketed into deep outer space to see God at the behest of a ravishing female alien from the planet Mingor.

Chuckling, he knew that would make people think he was a raving lunatic. Even though it had actually happened, he had come to realize that Sam was right; no one would believe him.

Forcing himself to get real, he settled at the keyboard and decided to opine about how the Davie rodeo grounds needed an upgrade and –

The doorbell rang. He guessed it was UPS as he had recently ordered some new running shoes online. Padding to the front door, he peered through the peek hole. His jaw dropped.

For there was Gina, gazing back at him. Although fearing his mind was playing tricks on him, he swung open the door.

Her lips broke into a soft smile. "I assure you, I'm for real, Geo."

She circled her arms around his neck and they hugged for several seconds before he stood back and studied her. She was clad in her biker chick get-up, the tight jeans, halter top and black boots that came up to her knees. Her hair fell over her shoulders, her face as gorgeous as that of a fashion model.

"Where's your ship? In the Everglades again?"

She turned and pointed. "In your parking lot. Hope I'm not disrupting the neighbors."

Stepping onto the front porch, he saw the shuttle right there, looking bigger than he remembered, towering above nearby parked vehicles.

It was already drawing a crowd of neighbors and onlookers. Traffic on the roadway in front of his townhouse was jammed with rubberneckers. He suspected the police would soon arrive.

He smiled. "Come in."

After he closed the door behind her, she answered another question before he could ask.

"I came back to be with you. I want to live here in Davie, Florida, in your townhouse – *with you.*"

Though elated to hear this, he was still skeptical. "You told me in so many words you don't love me and that you plan to connect with other men."

"I lied. I don't want to connect with anyone but you. I fell in love with you the first day we met at the park. I didn't realize how much I love you – until I left you."

He wanted to believe her. "How can I be sure that you won't leave me again?"

She came close. "You're going to have to trust me."

They were still in his bed, cuddling and caressing, when the news media and authorities arrived in force. Pulling on bathrobes, Gina and Geo went downstairs and looked out the front windows. The parking lot had turned into a sea of flashing lights, news vans and astonished bystanders, all gaping at the shuttle.

"Guess I should move the ship somewhere less intrusive," Gina said. "Maybe back out to the Everglades."

"Good idea. But I have a feeling the feds are going to want to talk to you first."

"This time, I'll tell them the truth about who I am, where I come from."

"I doubt they'll believe you."

"Oh, they'll believe me. If they don't, I'll give them the same treatment I gave those bikers."

"Should be interesting."

She rested her chin on his shoulder as they listened to the doorbell incessantly ringing. "There's another reason I came back to you."

"Let me guess. So that I can help you raise our daughter? She's due in fourteen months."

"Exactly. And, just as your horrible dream told you, she really will be smarter than me."

"You know about my dream?"

"Of course. It's still in your head. Don't worry. The pirates don't like to grill humans. They prefer to eat them raw."

"That's comforting."

He finally opened the front door and just in time. A S.W.A.T. team was ready to storm in as police had obtained a search warrant. Behind them were scores of federal agents from dozens of agencies. Among them were Bud Kowalski and Bob Klug.

Geo greeted the two men warmly. "Bud, Bob. How you boys doing this fine morning?"

Face perplexed, Kowalski jerked a thumb over his shoulder at the space shuttle. "What the hell is *that* doing in your parking lot?"

"I think I'll let Gina answer that."

The two stepped out on his front porch. The crowd of law officers, reporters and bystanders went quiet. Video cameras and smartphones snapped on.

"Folks, this is Gina," Geo said, putting an arm around her shoulder. "She comes from the planet Mingor. And that's her spaceship over there."

A few television reporters dropped their microphones, which, until then, had been thrust in Geo and Gina's direction.

Gina waved at the crowd. "Hello."

It took several hours for Geo and Gina – but primarily Gina – to convince authorities and media alike that she was, indeed an extraterrestrial with no malintent. By the time she was done answering all their questions, federal officials saw no reason to take her into custody.

Eventually all the commotion died down. The police cars, media vans and crowds dispersed. Geo and Gina were finally alone in his townhouse.

Geo went to his home office and sat down at his computer. Gina followed him and asked, "Now what?"

He punched up his story detailing their trip to see God, the one he had written for himself and the one Sam had rebuffed.

"Before you showed up, I didn't know what I was going to write about for my weekly column. Now I have an exclusive. I'll just run this thing."

She grinned. "Sounds good."

39

The wedding was simple yet elegant, held on a Sunday afternoon in mid-May on Hollywood's beach. A wooden-arch altar, bedecked with yellow daffodils, was set up thirty yards from the surf.

A quartet of musicians, playing violins and violas, sitting on white folding chairs, provided background music. A series of torches atop poles were positioned around the wedding party.

Gina and Geo stood barefoot in the sand. Both were dressed in shirts and jeans. She also wore a white veil, partially covering her face.

"Dear friends, family, Earthlings and residents of Mingor, we are gathered here today to witness the union of George Leon and Gina Martinez, a marriage that is, quite literally, out of this world," commenced Saul Finkelstein, who presided over the ceremony.

That drew a wave of applause.

"Though they haven't known each other all that long," Finkelstein continued, "Geo and Gina realize their hopes and dreams are more meaningful together rather than apart. For them, a long-distance relationship never would have worked considering their two planets are separated by billions of miles!"

More cheers and laughter. Gina winked at Geo. *This guy is a hoot*, she said telepathically. *He's perfect for the occasion*.

Agreed, Geo communicated back. *I'll have to buy him a beer after all this.*

Finkelstein paused for dramatic effect. Behind him, ocean waves lapped to shore. In front of him was a huge bouquet of microphones, which fed major networks from around the world.

"So, now," Finkelstein continued, "Geo and Gina will create a new family, one that includes all those here today and the millions watching around the globe. Their union commences a new era wherein we bond with new outer-space neighbors and try to find peace among all the peoples of the Earth."

That drew cheers.

The wedding party included fifty super-advanced humans, all women clad in white jumpsuits, from Mingor. Joining them were twenty-five of Geo's friends, coworkers and relatives. Bud Kowalski of the FAA and Bob Klug of ICE sat in a back row.

Beyond the wedding party, more than thirty-thousand onlookers had squeezed into the beach area, coming from all corners of the planet. Most simply wanted to see real extraterrestrials.

"The most remarkable moment in life," Finkelstein went on, "is when you meet the person who makes you feel complete and makes the world a magical place. Despite coming from two different worlds, Gina and Geo have met that person. Theirs is a pure and wonderful merger."

Gina gazed at Geo with heavy eyes as television cameras zoomed in close on her. *I can't wait to become Mrs. George Leon.*

Me either.

I'm really looking forward to our wedding night.

Me too.

I'm going to pour ketchup all over you and eat you alive.

Geo chuckled, prompting TV commentators to speculate that he was overjoyed to be marrying a space alien. Finkelstein wondered what was so funny. Geo silently mouthed "sorry" to his friend.

Finkelstein continued. "Both Gina's and Geo's family and friends have come here today to offer love and support, as has the world beyond. I know Geo and Gina want to thank one and all for making their day so special, particularly those from Mingor, who flew across the universe to be here."

Applause, cheers.

What are they serving at the wedding dinner? Gina asked Geo telepathically.

Geo had reserved a ballroom of the nearby Margaritaville resort for the post-ceremony party.

I ordered your favorite meal.

Meatball subs?

Damn straight. I wasn't about to skimp on an occasion like this.

Thanks, Geo. You're a germ.

Gem. Not germ.

Sorry.

"Like all of you, I am in awe, given the extremely unusual nature of this wedding," Finkelstein said to his audience. "Yet this much holds true: It doesn't matter where we come from, if love prevails. Gina and Geo's love will not only nourish each other but also the two worlds from which they come."

Cheers, applause, adulation.

You never have to worry about me leaving you, Gina relayed to Geo as they gazed at each other on the beach, standing under the arch.

At least until I'm dead.

I'm being serious here, Geo.

Don't worry. I'm not planning to die anytime soon.

Good. We've got a daughter to raise together.

Although some tabloids were already speculating that she was pregnant, Geo and Gina wanted to keep their unborn child a secret for as long as possible. They didn't want to give the media another reason to hound them. Both were already under the white-hot glare of the public spotlight.

Finkelstein's voice rose. "We, their community of friends and family, have a responsibility to this couple. With steadfast care, respect and love, we can support their marriage and the new family they are creating today."

That drew a chorus of "amens" and "hell yeahs."

Although hundreds of law enforcers were on hand, authorities didn't expect any trouble largely because of Geo's column. Gracing the front page of the *West Broward Banner* a week earlier, it consisted of his original story embellished with a plea for humanity to reach the next level of civility. The column was picked up by media worldwide.

I want to name our daughter Alicia, Gina said to Geo as the world soaked in the ceremony. It also was being broadcast on Mingor by way of super-advanced technology.

Does that name have some sort of celestial meaning?

No. It is in honor of your mother, Alice. She's a very nice woman.

Geo glanced at his mother, who was sitting next to his brother. She gave him a subtle wave and he subtly nodded back.

I really like your mom, too, Geo said to Gina. *Now I see why you're so beautiful.*

Gina responded with a grin. Her mother's contingent had arrived three days prior to the wedding. Flying in perfect formation, their ten shuttles had glided silently across the South Florida skies before landing at Fort Lauderdale-Hollywood International Airport, creating a frenzy of traffic accidents and jams below.

The contingent then checked into the swanky Westin Resort and Spa on Hollywood beach. There the aliens indulged in fine dining, swam in the ocean and hung out at the pool, all courtesy of the hotel's management.

Finkelstein raised his arms. "If you're not already standing, all please rise." The wedding party as well as thousands of others in the beach area, including those in bars and restaurants, got to their feet.

"Geo, Gina, I ask that you two remember that love is rooted in faith and trust. No other ties are more tender, no other vows more sacred than those you will now assume."

People from one end of the beach to the other went deathly silent to hear the best part of the ceremony. Those from Mingor held hands, which was shown on television. Suddenly everyone in the beachside crowd held hands with the person next to him or her.

"Do you, George Leon, take this woman to be your lawfully wedded wife, to have and to hold from this day forward, for richer, for poorer, in sickness and in health, to love and to cherish till death do you part?"

As they faced each other, Geo resolutely said, "I do."

When it was her turn, Gina telepathically said, *I do.*

Finkelstein's face crunched into a question mark. *Out loud, please,* Geo advised her.

Oh, sorry. "I do," she declared.

"You may now kiss the bride."

Geo took her in his arms, pushed back her veil and stared into her lovely eyes. As their lips met, the applause and cheers were deafening, not only at the beach but across the world.

About the Author

Ken Kaye has authored eight novels, *The Alien Biker Chick* being his latest. A graduate of the University of Denver, Ken worked four years at the *Sun Newspapers* in Northeast Ohio and thirty-seven at the *South Florida Sun Sentinel* as a reporter, editor and columnist.

Ken lives in Weston, Florida, with his wife, Maria.

Other books by Ken Kaye, available at Amazon.com:

The East Side of Lauderdale
Final Revenge
Stuck on 75
Gash in the Glades
The Wrong Hangar
The Kiss and Kill Girl
The Monroe Massacre

Made in the USA
Monee, IL
08 November 2026

68012209R00132